Transparency for Sustainability in the Food Chain

Transparency for Sustainability in the Food Chain

Challenges and Research Needs—EFFoST Critical Reviews #2

Gerhard Schiefer and Jivka Deiters

AMSTERDAM • BOSTON • HEIDELBERG • LONDON
NEW YORK • OXFORD • PARIS • SAN DIEGO
SAN FRANCISCO • SINGAPORE • SYDNEY • TOKYO
Academic Press is an imprint of Elsevier

Academic Press is an imprint of Elsevier
The Boulevard, Langford Lane, Kidlington, Oxford, OX5 1GB, UK
225 Wyman Street, Waltham, MA 02451, USA

First published 2013

British Library Cataloguing-in-Publication Data
A catalogue record for this book is available from the British Library

Library of Congress Cataloging-in-Publication Data
A catalog record for this book is available from the Library of Congress

ISBN: 978-0-12-417195-4

For information on all Academic Press publications
visit our website at **store.elsevier.com**

This book has been manufactured using Print On Demand technology. Each copy is produced to order and is limited to black ink. The online version of this book will show color figures where appropriate.

CONTENTS

LIST OF CONTRIBUTORS

1. Introduction

Gerhard Schiefer, University of Bonn, Germany
Jivka Deiters, University of Bonn, Germany

2. Problem Scenario and Vision

Gerhard Schiefer, University of Bonn, Germany
András Sebök, Campden BRI Magyarország Nonprofit Kft (CCH), Hungary

3. Stakeholders for Transparency

3.1 The Role of Government and Public Policies

David Barling, Centre for Food Policy, City University London (City), United Kingdom
Katarina Lorentzon, The Swedish Institute for Food and Biotechnology AB (SIK), Sweden
John Erik Hermansen, Faculty of Agricultural Sciences, University of Aarhus (AU), Denmark

3.2 Consumers and Society

George Chryssochoidis, RLabs Market Research Ltd., Greece

3.3 Agriculture as Stakeholder in the Chain

Daniel Martini, Kuratorium für Technik und Bauwesen in der Landwirtschaft e. V. (KTBL)

3.4 The Food Processing Industry

Tim Hogg, The European Association for Food Safety (SAFE), Belgium

3.5 The Role of Retail

Gerhard Schiefer, University of Bonn, Germany
Jivka Deiters, University of Bonn, Germany

4. Challenges and Experiences

4.1 Transparency Challenge 1: Transparency for Trust in Food Safety

Harmen Hofstra, The European Association for Food Safety (SAFE), Belgium
Tim Hogg, The European Association for Food Safety (SAFE), Belgium

4.2 Transparency Challenge 2: Transparency for Trust in Food Quality

Dietrich Knorr, Department of Food Biotechnology and Food Process Engineering, TU Berlin, Germany
Henry Jäger, Department of Food Biotechnology and Food Process Engineering, TU Berlin, Germany
Björn Surowsky, Department of Food Biotechnology and Food Process Engineering, TU Berlin, Germany

4.3 Transparency Challenge 3: Transparency for Trust in Food Chain Integrity

Karin Östergren, The Swedish Institute for Food and Biotechnology AB (SIK), Sweden
David Barling, Centre for Food Policy, City University London (City), United Kingdom

John Erik Hermansen, Faculty of Agricultural Sciences, University of Aarhus (AU), Denmark
Niels Halberg, International Centre for Research in Organic Food Systems (ICROFS), Denmark
Ulf Sonesson, The Swedish Institute for Food and Biotechnology AB (SIK), Sweden
Donna Simpson, Centre for Food Policy, City University London (City), United Kingdom
Katarina Lorentzon, The Swedish Institute for Food and Biotechnology AB (SIK), Sweden
Lizzie Melby Jespersen, International Centre for Research in Organic Food Systems (ICROFS), Denmark

4.4 Transparency Challenge 4: Signaling Information to Build Confidence and Trust in the Food Chain
David Barling, Centre for Food Policy, City University London (City), United Kingdom
Donna Simpson, Centre for Food Policy, City University London (City), United Kingdom
George Chryssochoidis, RLabs Market Research Ltd. (RLabs), Greece
Olga Kehagia, RLabs Market Research Ltd. (RLabs), Greece

4.5 Transparency Challenge 5: Technological Baseline Infrastructure for Tracking and Tracing
Daniel Martini, Kuratorium für Technik und Bauwesen in der Landwirtschaft e. V. (KTBL), Germany
Martin Kunisch, Kuratorium für Technik und Bauwesen in der Landwirtschaft e. V. (KTBL), Germany

5. Good Practice Experiences: An Integrated View
Xavier Gellynck, Department of Agricultural Economics, Ghent University (UGent), Belgium
András Sebök, Campden BRI Magyarország Nonprofit Kft (CCH), Hungary
Adrienn Molnar, Department of Agricultural Economics, Ghent University (UGent), Belgium
Attia Berczeli, Campden BRI Magyarország Nonprofit Kft (CCH), Hungary
Katrien van Lembergen, Department of Agricultural Economics, Ghent University (UGent), Belgium
Miroslav Bozic, Campden BRI Magyarország Nonprofit Kft (CCH), Hungary
Fruzsina Homolka, Campden BRI Magyarország Nonprofit Kft (CCH), Hungary

6. Enabling Activities and Needs for Action

6.1 Transparency Challenge 7: Communication with Stakeholders and Media
András Sebök, Campden BRI Magyarország Nonprofit Kft (CCH), Hungary

6.2 Transparency Challenge 8: Dealing with Claims and Data Ownership
Gerhard Schiefer, University of Bonn, Germany
Jivka Deiters, University of Bonn, Germany

6.3 Transparency Challenge 9: Coordination and Cooperation Initiatives
Gerhard Schiefer, University of Bonn, Germany
Jivka Deiters, University of Bonn, Germany

7. Concluding Remarks
Gerhard Schiefer, University of Bonn, Germany
Jivka Deiters, University of Bonn, Germany

ACKNOWLEDGMENT

This is a publication developed with support by the European Union within a dedicated project initiative

Contract No.: FP7-KBBE-2009-245003
Coordination and Support Action—CSA;
Food Quality and Safety
Transparent_Food
Quality and integrity in food: a challenge for chain communication and transparency research

This research has been supported by the European Commission through its FP7 program and the project "Transparent_Food" (*Quality and integrity in food: a challenge for chain communication and transparency research*), project/contract number 245003.

EXECUTIVE SUMMARY

Transparency in the food sector and especially toward consumers is one of the priority issues on the agenda of consumer policy and consumer representatives. Food scandals and deficiencies in consumer communication have raised consumers' requests for information. The requests for improved transparency are also due to increasing interests of consumers and policy in food that is not only safe, and of the quality they expect, but also matches evolving expectations that food production is based on processes that limit negative impacts on the environment and consider social concerns.

Moving toward improved transparency requires action by stakeholders of the food chain as well as knowledge of where and how to move. The Strategic Research Agenda of the industry-led European Technology Platform "Food for Life" that was published some years ago identified deficiencies in our knowledge on transparency and asked for research initiatives that were, however, not further specified.

This book aims at specifying the research needs that could, if followed, facilitate developments toward transparency. The results will be presented in a hierarchical view, starting with goals and continues with major research challenges and expected deliverables. The goals being addressed are listed below. They give a first indication on the issues covered in the chapters. While priorities are difficult to identify, the first goal in each domain and, consequently, the first research challenge linked to goals, represent some priority understanding. The results are based on literature reviews, case studies, surveys, web consultations, stakeholder workshops, scientific conferences, expert workshops, and participants' experiences. Literature reviews are captured in separate reports, which constitute the reference list of this book.

The presentation introduces into the subject through a first view on the relevance of the discussion for different stakeholders (Chapter 2). This is followed in Chapter 3 by a detailed discussion on the state-of-the-art and on research challenges in a number of specific domains that represent major directions of discussion and are linked to specific research domains. They involve food safety, food quality, chain integrity, the link with consumers, and the technological base of tracking and tracing systems. These discussions are confronted in Chapter 4 with conclusions that evolve from comprehensive best practice studies representing the experience of the industry. The discussions related to specific domains and best practice experiences constitute the major thrust of analysis in the compendium. As the discussions have a focus on the total food chain, different research domains might develop views with some overlap. They are accepted in this publication as they facilitate the identification with the specific discussions going on in the selected research domains.

The presentations are complemented by discussions of research needs in activity domains with relevance for system development and implementation. They deal with issues around consumer communication, the utilization and substantiation of claims, the consideration of data ownership, information markets, the coordination of sector initiatives, and the support in developing sector capabilities.

SUMMARY—LIST OF GOALS

Food Safety: Challenge 1

Goal 1: Addressing transparency issues related to emerging food safety risks

Goal 2: Ensuring that transparency does not impede emerging technologies from achieving their potential

Goal 3: Providing a fair and functional governance of food safety

Goal 4: Understanding the effects of the parallel economy on food safety

Food Quality: Challenge 2

Goal 1: Food chain—better integration from farm to fork

Goal 2: Traditional and emerging technologies: A synchronized assessment

Goal 3: Analytical methods: Improving speed, detection limits, and process adaptation

Goal 4: Improving food quality standards and making provisions more stringent

Chain Integrity: Challenge 3

Goal 1: Valid indicators for estimating the integrity performance within an operational and sound traceability reference unit

Goal 2: Cost-effective systems for data collection and sharing that take advantage of existing data collected through a food chain
Goal 3: Robust concepts for guaranteeing the integrity performance of different food chains

Signals for Communication: Challenge 4

Goal 1: A more sustainable food chain that utilizes transparency in signaling its sustainability criteria from business to business and on to the consumer
Goal 2: Providing signals around the environmental, social, and ethical aspects of food that are understood by consumers and respond to their needs
Goal 3: Establishing consumer trust (the role of the media) and managing the transition to greater transparency
Goal 4: The development and utilization of technologies to facilitate the flows of information and transmission of signals thus enabling better transparency

Technology and Tracking/Tracing: Challenge 5

Goal 1: Making different subdomain-level data encodings interoperate
Goal 2: Feasible identification of holdings, production sites and units, and sound definition of traceability reference units in primary production
Goal 3: Supporting balancing of demands for confidentiality versus demands of open information
Goal 4: Sector-wide economic and technical feasibility of a baseline information infrastructure

Best Practice: Challenge 6

Goal 1: Developing optimal transparency systems
Goal 2: Understanding cost and benefits of transparency systems
Goal 3: Creating multitarget transparency systems
Goal 4: Identifying best practice transparency systems as reference systems for future scenarios

Communication: Challenge 7

Goal 1: Improving the access of stakeholders to transparency information

Goal 2: Organizational specification of efficient and balanced transparency systems with fitting levels of detail
Goal 3: Improving the exchange of transparency information between consumers and small/medium-sized enterprises
Goal 4: Establishing open innovation exchange between consumers and members of the chain at various stages of the chain

Claims and Data Ownership: Challenge 8

Goal 1: Substantiation of claims
Goal 2: Protecting and considering data ownership

Coordination and Cooperation Initiatives: Challenge 9

Goal 1: Identifying suitable organization infrastructures for coordination support
Goal 2: Reaching a sector status in information availability and handling that fits needs
Goal 3: Designing markets for information and claims

CHAPTER 1

Introduction

This Research Agenda on "Transparency in the Food Chain" goes back to the Strategic Research Agenda of the European Technology Platform "Food for Life" (http://etp.ciaa.be) where "*transparency in the food chain*" was mentioned as one of the priority areas for competitiveness of the European food sector that required dedicated research initiatives. This initiated the project "*Transparent_Food*" (quality and integrity in food: a challenge for chain communication and transparency research; www.transparentfood.eu) which aims at the identification of barriers and research needs toward improvements in transparency. The project received funding through the EU-Commission within its Seventh Framework program and builds on the engagement of a European consortium with close relationships to working groups of the European Technology Platform "Food for Life." An advisory board with representatives of stakeholder groups provided support.

Transparency is an emerging issue which depends on innovations in organization and communication. This marks a shift in focus as prime manifestations of innovativeness by food organizations are traditionally linked to product and process innovations.

Although process innovations may be defined as new tools, devices, or procedures, as well as knowledge in throughput technology that mediate between inputs and outputs, product innovation may be seen to do more with the outputs that are introduced for the benefit of consumers and citizens. A plethora of factors has been associated with product innovation. Yet, transparency aspects have not only acquired visibility, they indeed start shaping up entire domains of such influencing factors.

Consider, for instance, country-level environmental factors impacting upon product innovations. Transparency discrepancy-related institutional differences among economies are likely to lead to substantial variation in innovations strategies and patterns of innovative performance.

Consider sector- and business-level external environments. Rapid environmental change and the uncertainty that this often creates for food organizations' decision-makers certainly stimulates innovation. Product innovations are most prevalent and useful in uncertain environments in which competing products or consumers' preferences alter significantly. Consumer and food chain requests for fast and greater transparency will almost certainly increase speed and complexity and thus oblige whole business sectors in forward leapfrog-like transparency-related actions for survival.

Consider also how competitive dynamics and hostility will be affected by requests for faster and greater transparency. Food product innovation activity may likely be strongly and positively affected by such competitive dynamics. Perceived increased hostile competitive environments due to increased transparency will also place further pressure on food firms to initiate product innovations. These innovations in effort will improve their ability to make the best use of the resources they have in order to satisfy transparency issues. Increased spending on R&D will also result as food firms will have to quickly copy each other and preempt competition.

Finally, food firms' market orientation will be affected. Fulfilling requests for greater transparency is likely to oblige further adoption of the marketing concept philosophy and foster the development and marketing of radically and probably better new products, as well as important product modifications.

Transparency itself is a fuzzy domain which very much depends on the perception of people, their background, cultural environment, situation, and expectations. One of the challenges in the sector is to reach a level which can be accepted by a majority of stakeholders as sufficient. It is not only the focus of the project to identify what is sufficient but also to identify research activities that could contribute to reaching an understanding on:

a. what could be considered as present or future "best practice" and
b. what are knowledge deficiencies that limit developments toward this status and require research to overcome.

It is obvious that research alone cannot provide transparency. It requires the preparedness of actors in the field, including policy, enterprises, and service operators to act. Furthermore, as transparency builds on information from throughout the food chain (including agriculture), action requires cooperation and coordination.

In a sector with an open network situation and a majority of small- and medium-sized enterprises (SMEs), a coordinated development path is difficult to reach. One might argue that transparency, if requested by consumers, will eventually be reached. However, this might take time if no dedicated coordination activities will be initiated. But whatever the approach, research can facilitate the development and allow a more focused development, reducing risks of failures.

Through collaborative efforts by leading experts from 11 universities and research institutes covering a wide range of research disciplines, the project has captured the present state of the art and deficiencies that required research activities in a number of extensive reports. The analysis involved the review of literature, projects, experiences, and communication with stakeholders through surveys and workshops. A specific initiative involved the analysis of "best practice" examples which demonstrate proven working levels of transparency.

The project summarizes the results in this "Research Compendium" to provide guidance in the initiation of new research projects that could support the sector's development toward better transparency. It identifies, in a compressed approach, the state of the art, goals, research challenges, and expected deliverables. In Chapter 3, the focus is on tracking and tracing schemes, the information domains of food safety, food quality, food integrity, and the communication with stakeholders through signals and messages. It is complemented by an analysis of experiences from best practice research (Chapter 4) and research proposals linked to communication with stakeholders, the consideration of claims and data ownership and the coordination of sector activities (Chapter 5).

CHAPTER 2

Problem Scenario and Vision

Transparency is driven by needs. The vision is to reach transparency for everybody from whatever background and in whatever situation and perception. Reaching this state is, however, a challenging and ongoing task considering the dynamics in scenarios and needs.

> Consumers' trust in food, food production, the origin of food, and the actors involved is a core requirement for the functioning of European food markets and the competitiveness of industry. With the experience of the BSE crises and subsequent food scandals in mind, consumers increasingly expect transparency on which trust can build. Transparency is not meant to know everything but to *create awareness* on the issues that consumers and customers in the chain are interested in, involving information on the safety and quality of products and processes, and increasingly on issues around environmental, social, and ethical aspects.

Transparency involves the process of providing transparency (*process-based aspect*) while at the same time considering the balance of interests between recipients and providers (*power balance aspect*) involving consumers and all stakeholders in the chain:

1. *Process-based aspect*: Transparency is a set of measures for building up credibility for consumers and customers, through openness, trust, and accountability on activities along the food chain, by underpinning the verity of messages, and by generating the perception of being informed to allow (informed) decisions. This is achieved by making appropriate signals/information available and understandable on the verity of messages (claims, statements) on specific characteristics of products, processes, production environments, activities of actors, and the cultural and legal background of the production which cannot be substantiated by the usual quick and simple methods, characterized either of a positive enhancement or of a negative risk reduction nature.
2. *Power balance aspect*: The realization of transparency builds on the consideration of the valid, or perceived as such, needs of consumers or

customers for facilitating their noninformed or informed decisions and the sound balance with confidentiality needs of food chain members as providers of transparency.

Transparency is one of the most complex and fuzzy issues the sector is facing. The complexities are not only due to complexities in food products and processes but also due to the dynamically changing open network organization of the food sector with its multitude of SMEs, its cultural diversity, its differences in expectations, its differences in the ability to serve transparency needs, and its lack of a consistent appropriate institutional infrastructure that could support coordinated initiatives toward higher levels of transparency throughout the food value chain and on a global scale.

Transparency builds on appropriate *signals* that integrate available *information* and communicate a certain "*message*" to recipients (e.g., food is safe). Transforming information to simple, clear, and easily understandable "messages" and ensuring that messages build on information that can be trusted are key issues in ensuring transparency and trust. The provision of information could involve a broad range of alternatives depending on opportunities but also on the ability and willingness of consumers and decision makers to grasp, interpret, and process the information as needed. Transparency does not build on the communication of an ever-increasing number of information items. The sector has developed many approaches for suitable aggregations and certifications along the way.

Examples include "*carbon footprint calculations*" aggregating various emissions of greenhouse gases or "*GlobalG.A.P. certificates*" building on the fulfillment of a broad range of requirements related to food safety, food quality, or environmental and social concerns. The knowledge about the content of certificates allows for disassembling the information bundle into individual information items at the users' end if necessary. However, in daily life, users are probably more interested in the "message" ("this food is safe") than in the detailed background information as such, but must at the same time be able to have access to trustworthy information for developing and keeping trust.

Trust is a sensitive "product." Non-government organisations (NGOs) and other groups are increasingly demonstrating the divergence between claims and reality contributing to the public push toward increased transparency. Claims in this context represent statements on product characteristics (e.g., on quality) that are not directly apparent to consumers upon visual inspection. However, with the scattered company infrastructure of the sector, the deep integration within the food industry and the importance of commodity products in most food products across the sector, reaching transparency is a sector problem that cannot be tackled by individual companies alone. Furthermore, any scandal is damaging the trust in "food" in general, distorts markets, and cannot be limited to the individual company involved.

This requires sector-wide efforts to improve transparency linked to sales products. However, with the dependency of transparency information on the activities of all actors in the value chain, the design of appropriate transparency systems requires cooperation within the sector and a suitable IT infrastructure on which information can be collected, processed, and moved toward the retailer and the consumer. The IT infrastructure is the critical success factor in the scenario as, without its base, any further agreements in the sector on the development of transparency is without a realistic chance of implementation.

The baseline for such an infrastructure is the ability to clearly identify products, the link between products of varying identity (e.g., levels of processing) at subsequent stages of the food value chain, and the transparency information. This so-called *tracking and tracing ability* is the base on which all information and services can build.

When trying to capture an overview on research, project and industry initiatives linked to "transparency" and the needs for further initiatives and research, we may constitute that a good source of knowledge and experiences is available. However, this knowledge is either not documented or usually separated into many diverse domains defined not only by subject areas (e.g., information science and management science) but also by actor groups (e.g., research, industry, and policy) with little communication among them.

CHAPTER 3

Stakeholders for Transparency

Stakeholders for transparency involve a diverse group making it obvious that any transparency solution would need to build on a balanced consideration of different transparency interests.

The consideration of stakeholders needs to take into account the stakeholders at the different stages of the food value chain. This encompasses a broad range of enterprises engaged as suppliers to farms, as farms, as enterprises in processing and trade or retail, and consumers as the final customer in the chain. The complexity is exemplified in Figure 3.1. The situation is further aggravated by the dynamics in trade relationships that constitute an open network scenario with changing trade relationships due to variations in the quantity and quality of agricultural production.

There is general agreement that information for reaching transparency may involve information on economic (including food quality aspects), environmental (including carbon emissions and similar aspects), social (including food safety aspects), and ethical concerns (including fair trade or animal welfare aspects). Differences between stakeholders focus on interests within these domains, the appreciation of the domains, and communication to customers.

This allows research initiatives to consider a range of suitable transparency issues and to leave it to systems that deliver transparency to dedicated stakeholders to decide on priorities, content, and communication.

Apart from individual stakeholders within the food value chain, *society* as a whole might be interested in information about issues that affect society irrespective of people's role as consumers or enterprise actors. Food safety, environmental issues, and social concerns are primarily of interest. Society in this respect is represented by policy which

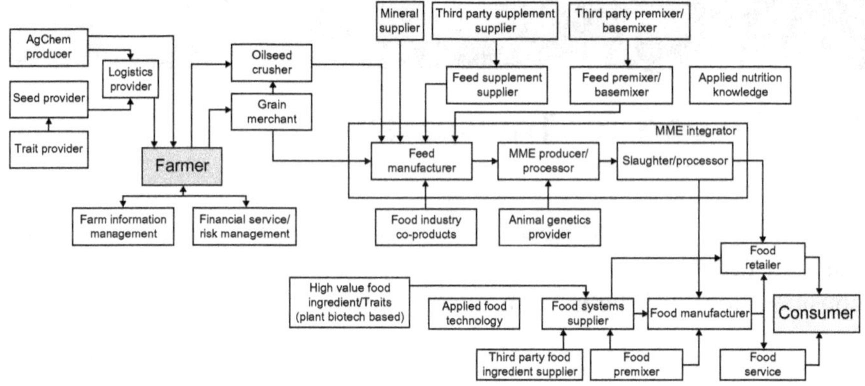

Figure 3.1 Complexity of the food chain – an illustrative example.

may provide guidance, incentives, or impose regulations to assure society's interests are being followed.

With a view on individual stakeholders, *consumers'* interest in transparency is of foremost relevance considering their role as the recipients of the final product of the chain who determine the market success of any chain activity. *Enterprises* that serve their markets need to know consumers' interest in their procurement and production decisions. If consumers are interested in a certain product characteristic and request information, enterprises all along the chain that can influence the characteristic depend on receiving the appropriate information. This involves retail and industry but may also reach back to agriculture and even to the input industry.

However, the realization of transparency is not just an issue of information collection and provision but depends on a balanced consideration of interest. While the provision of transparency may create benefits to recipients, its realization depends on information that needs to be collected, processed, and prepared for provision, activities that create costs.

To sum it up, the long-term sustainability of the food sector depends on a *balanced consideration* of the different interests which requires the sector to develop appropriate schemes for information exchange (e.g., *information markets*) and for the realization of transparency.

Furthermore, providing information to other stakeholders may have negative impacts on competitive advantage of the initial owner of information. As a consequence, stakeholders who provide information and incur the costs are, in principle, only prepared to share information with other stakeholders if they perceive more advantages than disadvantages. The power structure in the value chain might enforce information owners to provide information even against their own perceived interests; a situation that is sometimes related to agriculture.

3.1 THE ROLE OF GOVERNMENT AND PUBLIC POLICIES

The transmission of appropriate communication and signals from food value chains and networks to the consumer, based upon clear transparency and traceability, is an integral element in establishing and maintaining the European public's trust in food—where the public exists as both consumer and citizen. In addition, these elements are prerequisites for the effective working of the European market. The lack of appropriate information transmission and so the breakdown in risk management over the food safety crises in the 1990s led to governments stepping in and a wave of reforms that instituted national Food Safety Agencies across Europe, as well as the European Food Safety Agency and the General Principles of EU Food Law. The General Principles legally embedded traceability into food chain management responsibilities.

Traceability demands the effective transmission of information along food chains, but transparency is another feature increasingly required, to meet more specific societal and public demands for more knowledge about the processes by which the food product is produced. This knowledge, in turn, adds value to the product, in addition to the intrinsic characteristics of the product. Food certification and labeling can further public policy goals across areas such as environmental and natural resource protection (e.g., biodiversity impact and integrated farm management and organic farming processes; sustainable fisheries; plus growing interest in embedded water and water stewardship), climate change and energy use (carbon labels, food miles), food safety (food hygiene), public health and dietary goals (nutritional content and nutrient reformulation of products), and social and ethical issues (e.g., labor standards, fair trade, animal welfare, geographical origins).

However, the public policy makers and regulatory agencies are increasingly reliant upon the effective actions of the food supply chain to deliver effective standards as to the origins and qualities of the food produced, in both product and process terms, in order to realize the regulatory objectives of EU law. Hence, the governance of food is a highly complex and symbiotic area of public and private knowledge interchanges in which transparency is a key feature.

Policies shape and influence the ways that food chains and networks are formed and their performance. In food policy, the governance of food chains are shaped by the interaction at multigovernmental levels from international regimes (from the World Trade Organization (WTO) trade rules agreements to the Kyoto protocol) to local authorities, both through public policy regulation and the introduction of government-endorsed standards and policy instruments (including financial), exerting both negative and positive inducements.

Public policy interacts with private-sector-led governance, which can include: private corporations and companies, trade associations, professional bodies and expert (epistemic) communities, social enterprises, and nongovernmental and civil society organizations.

Private forms of governance are mobilized or utilized by public bodies to achieve desired public policy goals and involve policy networks of actors and institutions. In contemporary food chains, for example, food product process-related characteristics are being enhanced through private-led certification schemes with the public policy makers endorsing, implicitly or explicitly (through regulation), the goals and activities of these schemes in areas beyond food safety assurance.

Food and feed production and consumption contribute substantially to the environmental footprints of society. According to the Environmental Impacts of Products (EIPRO) study published by the Joint Research Center of the European Commission in 2006, food and drink, together with transport and housing, are consistently the most important areas responsible for the greatest environmental impacts in the EU—across different studies and the different impact categories.

Together they account for 70–80% of the whole life cycle impact of products. These concerns, together with ethical and social concerns, are reflected in a number of policy measures nationwide and within EU on the one hand; and in actions from consumer groups, NGOs, and companies on the other hand. Major aspects on the agenda at present are emissions of greenhouse gases contributing to global warming, loss of biodiversity and wildlife through intensive agriculture and fisheries, exploitation of poor farmers and farm workers, unfair sharing of profits along the chain, and animal welfare issues.

The foresight analysis "Sustainable food consumption and production in a resource-constrained world" presented in February 2011 by the Standing Committee on Agricultural Research (SCAR) presented a set of principles on which food chains should be based in order to meet the challenges of the future are presented. These include: (1) high quality of life of all stakeholders involved in food systems from producer to consumers, (2) reducing our footprints related to food consumption, and (3) resource conservation to avoid irreversible loss of natural resources.

Likewise the Organisation for Economic Co-operation and Development (OECD) concluded in its preliminary report "A Green Growth Strategy for Food and Agriculture" (June 2011) that a major shift in farm policy and practice is needed if a growing world population is to be fed without overexploiting scarce natural resources or further damaging the environment. One of the three priority areas for coherent action is to ensure that well-functioning markets provide the right signals: "Prices that reflect the scarcity value of natural resources as well as the environmental impact of farming will contribute to greater efficiency. Economically- and environmentally-harmful subsidies should be phased out. The *polluter pays* principle needs to be enforced through charges and regulations. Incentives should be provided for maintaining biodiversity and environmental services" (OECD, 2011). Internalizing environmental costs and "getting the prices right" will be necessary to reach and maintain a *sustainable agriculture, food production, and food consumption.*

The rapidly-evolving policy area of sustainable food consumption and production (SCP) illustrates the complex and symbiotic nature of these public–private relationships and knowledge exchange. At the EU level, there are a range of developing initiatives to measure the sustainability impacts of food and drink products and their production process methods. The European Commission, under the grouping of Environment Directorate-General has set up the European Food Sustainable Consumption Roundtable, involving all of the key food chain related peak European trade associations, in order to develop a harmonized environmental framework methodology for food and drink product. The United Nations Environment Programme (UNEP) is a partner in this initiative, and this international agency has been developing a social life cycle product assessment approach. In addition, DG Environment is developing a harmonized framework methodology for the calculation of the environmental footprint of products. The Commission has been looking at the suitability of the potential extension of the EcoLabel to food products and has launched a long-term policy goal (with milestones) of a Roadmap to a Resource Efficient Europe. The EU's development of metrics that measure the impacts (environmental, social, and ethical) of food chains and their products requires greater and more appropriate methods of transparency along food chains.

3.2 CONSUMERS AND SOCIETY

Societal developments are fundamental in shaping society members' well-being but also the dynamics of the societal system(s) themselves. It is also well understood that societal needs and wants overlap with citizen needs and wants, but also with individual and organizational consumption needs and wants. However, a distinction appears to exist when monetary flows and exchanges occur in the course of satisfaction of such needs and wants, societal and market realities portray a picture where nonmonetary flows and exchanges are also present including for instance election choices to follow specific policy mandates by electorates, as well as population support for specific alternatives at local, regional, national, or international level.

These are particularly evident in the food, agriculture, environment, and assimilated domains. Substantial imbalances in information availability and exchanges, systemic ineffectiveness and inefficiencies, are coupled with a perceived ever-increasing speed of changes in the environment, complexity, unpredictability, and instability. These imbalances drive society and its members to request an increasing ability to prevent, act, react, and balance through more complex governance (or self-governance) mechanisms but also to receive information for these purposes. The apparent logic for this is clear. In environments where speed of change is slow, characterized by low complexity, higher predictability, fewer perceived risks and higher stability, balance and strategy, as well as operational management of own affairs, is easier. In this case, systemic information can be limited, less detailed, less frequent, and more obscure. In environments where the reverse is true, either objectively or perceived, systemic information can often be obscure.

Such obscureness becomes increasingly articulated through multiple forms regarding societal matters but also consumer matters, irrespective of involvement of monetary or nonmonetary flows and exchanges. The required clarity in the food, agriculture, environment, and assimilated domains spans, as it has been discussed earlier, should be evident across a large number of foci and subjects ranging from ethics to quality to sustainability. Such requests do not also remain abstract in the sense that members of the society only express wishes which may not be satisfied. Such requests form attitudes, crystallize in beliefs, become specific intentions and prescribe policy mandates through electoral decisions, or when backed up with financial demand and monetary exchanges, dictate the works, shape up the nature of consumer and organizational markets, model economies, and influence international trade. Greater transparency, thus, following societal and consumer ordering becomes an overarching and ruling framework for societal and market operations guiding, directing and dimensioning strategies, and operational actions.

To sum it up, transparency is the basis for supporting consumers' in pursuing their interests in food that is safe, readily available, affordable, and of the quality and diversity they expect.

3.3 AGRICULTURE AS STAKEHOLDER IN THE CHAIN

Stakeholders in the chain involve agriculture, industry, and retail. While the transparency interests of industry and retail can be directly linked to consumers' interest, agriculture is different. It is on one side an important source of information required for consumer transparency, whilst on the other side, it is far away from consumers and might not directly benefit from increased engagements in the provision of transparency toward consumers.

According to EUROSTAT figures, in most countries across Europe, the number of farms is more or less evenly distributed over all size classes. Exceptions to this rule are several countries in the southern and south-eastern part of Europe (e.g., Greece and Hungary), where more than three quarters of the farms are small or very small. In general, but especially in smaller farms, the number of employees in farms is limited to a few, therefore not giving way to putting a lot of work force into supporting transparency. This is in contrast to documentation needs from a chain perspective. Farmers operate in a highly uncontrolled environment. They have to undertake a number of risky practices in food production leading to environmental impact or contaminations if not done properly.

For accumulating information that could serve transparency along the chain and toward consumers, farms have to be able to document relevant cultivation practices of their sales products. In recent years, growers have been obliged to comply with various requirements from policy or customers (e.g., cross compliance, good agricultural practice, and integrated crop management) as a precondition for access to subsidies and markets. In this documentation, environment, farmers, and even small farmers would principally be able to provide the information required for transparency if it could be realized without major investments and build on documentation and communication tools that are compatible with their existing documentation needs.

An important aspect for farmers is confidence that information provided to other stakeholders in the chain will not be used against their interests. For small farmers, organizational developments toward the involvement of *information trustees* might provide a solution. On a positive note, experiences show that quality farmers may be interested in providing information if it supports the appreciation of their contribution to food production with the final consumer product, at least by identifying the point of origin. It could assure that good traceability practices are in place that allow the identification of their products' destination and that their quality products have been handled appropriately.

Providing transparency is an issue which puts especially high pressure on farms which sets them apart from other stages of the food chain. Farms as the basis for all food production have to serve simultaneously different food chains which usually focus on specified groups of produce, for example, meat, vegetables, or cereals. They are furthermore confronted with differences in requirements regarding the provision of information in terms of content, format, scheduling, etc. Farms are therefore a bottleneck in the provision of transparency. Their appropriate integration into transparency schemes is a precondition for reaching transparency in food.

3.4 THE FOOD PROCESSING INDUSTRY

The European food industry is important by any measure. It is the largest manufacturing industry at the European level with a total turnover of almost 1 trillion Euros and directly employing 4.2 million people (Source: www.fooddrinkeurope.eu). The food industry is the major link between primary production and retail or foodservice (e.g., restaurants) for a large portion of the food produced and consumed in Europe. At present, it purchases and processes about 70% of the continent's agricultural production (source as above).

The industry has managed to adapt to the new regulatory environment that has developed since the publication of the EU's *White Paper on Food Safety* in 2000, and the profound organizational and technical changes that this has brought about. This is particularly notable when considering that more than 95% of the industry's 310 000 companies are SMEs (Source: www.fooddrinkeurope.eu).

From this brief description of the dimension of the industry, it becomes clear that it has a major role as participant and gatekeeper in many of the transparency issues described in this book. The food industry is the enabler of the majority of innovations in products and the initiator of many—this function is nowadays also shared with retail and other nonproducing enterprises such as brand holders. Many innovations involve alterations in practices that are not communicated to the end-user, such as alterations on formulation or processing that do not alter the essential quality of the final product. Other innovations are communicated to the final consumer with the intention of adding some value to the food or service sold. In the latter case, the elements of objective proof that support any message communicated to the final consumer will have to be transmitted down the supply chain in technical terms and then translated into the signals which are much debated here (see Figure 3.2).

Where innovations are not visible to the end consumer this information remains in technical terms. In cases where the product is

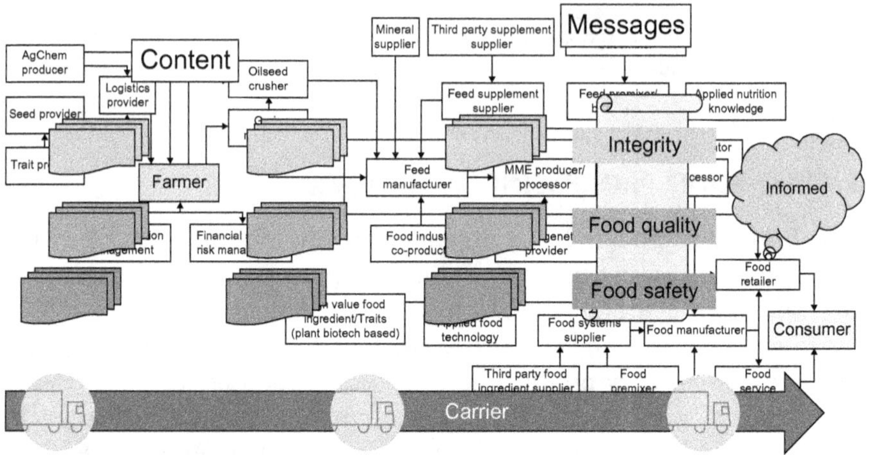

Figure 3.2 The flow of information along the chain and its transformation from data at initial stages to signals and messages at the end of the chain.

sold under a brand name which is not the producer's, then communication of all relevant technical details will be obligatory. In any case, the food industry, as the primary stage for integrating raw materials and ingredients and, in addition, secondary materials such as packaging, bears a significant burden of information integration. This involves costs that have to be integrated into the calculation of their production costs.

As suppliers to strong nonproducing brand holders, food companies might be subject to high pressures to comply with stringent demands which, while clear and objective in application, may not always be transparent to producers regarding their background. In such scenarios, the power imbalance between brand holder and supplier might be a potential threat to backward transparency, e.g., transparency toward suppliers. However, as discussed earlier, the long-term sustainability of the food sector and the development of sustainable trusted relationships depend on a *balanced consideration* of the different interests in the realization of transparency.

3.5 THE ROLE OF RETAIL

Retail constitutes the final stage in the food value chain which delivers goods to consumers and *provides* the information requested by the market, i.e., the consumers. Consumers on the other side link up with retail to receive goods and to *request* the information they need for making an "informed decision."

Understanding consumers' information needs is a difficult issue considering the diversity in backgrounds, customs, interests, etc. Retail as their direct business partner is in constant exchange with consumers and might be in the best position to understand its customers' needs. This does not mean that retail is always *able* or *willing* to provide to its customers all information they are interested in. However, retail is certainly interested in being always prepared, i.e., have all the information at its disposal which it might not provide in its daily routine but which customers might request with urgency in certain circumstances.

Looking at the interface from both sides, one can identify the following principal interests in dealing with needs in transparency.

The consumer perspective on transparency is primarily characterized by *pull situations* as information is usually requested. Examples of pull situations presently discussed in project developments include alternatives which build on passive or active consumer involvement. The *passive scenario* involves a consumer who is being warned in a shop if he/she intends to pick up a sales product that is not matching his/her profile. Various technology developments, for example smartphones, support these initiatives. In *active scenarios*, a consumer may enter a shop or approach product displays with a communication device that contains his/her profile. The device may guide the consumer (e.g., with support of shop maps) through the shop to the relevant product displays that match his/her profile.

Information to be provided to the consumers is usually directly linked to the sale of products and has to reach retail on a *routine basis* together with the product. Retail's interest in the assurance of food safety and quality is realized through its business procurement agreements specified at the time of order. As such, retail can expect to receive the products that match its orders. Any additional information is in principle *obsolete* and not part of the deliveries. However, retail might occasionally want to exercise some *control* involving requests for additional information or to being informed in time if, by any circumstances, shipped deliveries might not match (or no longer match) its order specifications in quality or safety.

Retail provides the interface between the value chain and the consumer as the final customer. As such it is interested to have all information at hand that the customer may ask for. Its strength is the ability to deliver, both in products and information.

However, beyond the ability to deliver information, retail must also assure that products are safe and of the quality its customers expect. It needs to receive the information that allows for eliminating unsafe products or products not matching its own or its customers' expectations. Furthermore, it needs it in time. However, there is a difference in information dependency.

Defining quality requirements at order times and requesting that suppliers assure that deliveries match requirements (which in turn could reduce communication needs) is a result of the power balance in the chain. With the strength of retail, suppliers have to perform and to deliver the product characteristics requested. The responsibility for analysis is placed on the early stages of the chain. In this case, information provided to consumers could remain quite stable and build on retail's order characteristics. The situation is different in scenarios where production is the dominant partner.

In this scenario, production could push products to retail and retail would need to analyze product(s) characteristics and to provide consumers with the ever-changing information about the actual product characteristics. In Europe, the first type of scenario is the dominant one which allows limiting communication needs between producers and retail.

CHAPTER 4

Challenges and Experiences

In the identification of transparency challenges evolving from a discrepancy between needs, state of the art, and experiences that will be discussed in the following part, the Strategic Research Agenda (SRA) has utilized a broad range of approaches, including literature analysis, best practice analysis, chain analysis, work group discussions, expert discussions, surveys, web consultations, and simulation studies to reach results that serve the objectives.

In its presentations, the SRA will primarily follow a "layer approach" that accounts for the complexity in transparency discussions. This is complemented by presentations based on an integrated view that includes *best practice experiences* from food value chains and some issues of *generic* nature.

In this chapter, the focus is on the layer approach, the integrated view follows thereafter. The layer approach distinguishes between:

a. *Upper levels* linked to the recipients of transparency, and
b. *Lower levels* linked to the actors in the food value chain and their production and distribution processes.

The different layers identify different communication needs (Figure 4.1). The lowest level provides the "*infrastructure*" for data communication. It is closely related to information technology and the identification of the path that a product is taking from production to consumption. This is linked to the tracking and tracing functionality which makes it feasible to communicate additional information as "backpack" on the tracking and tracing information base.

The next layer serves the *collection of information* about the various domains (food safety, food quality, chain integrity) of interest. This layer represents the classical information collection and communication approach. The third layer involves the *transformation of information into signals* or further to simple-to-understand messages like "this

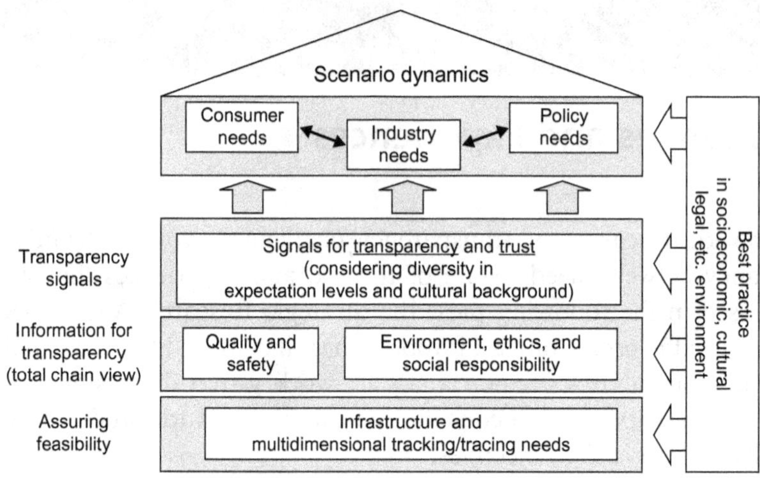

Figure 4.1 Multilevel structure of transparency challenges.

food is safe" which serve the transparency needs of the various stakeholders (consumers, enterprises, and policy) depending on the situation they are in (scenario).

In dealing with the information layers, the specification of information issues involves aspects such as "where or where from," "when," "what shape," "what guarantee or reliability," and "who owns." One needs to be aware that all layers build on a *total chain view* reaching from the source of production to the consumer as the final customer. In the chain analysis, this multilayer approach needs to be transferred into the total chain view where the collection of information, the preparation of signals, and the communication of signals depend on each other and are usually linked to different time periods and to different stages in the food value chain.

4.1 TRANSPARENCY CHALLENGE 1: TRANSPARENCY FOR TRUST IN FOOD SAFETY

4.1.1 Scope and State of the Art

The analysis of the current situation regarding food safety demands (regulations, commercially-applied specifications, consumer perceptions) is the basis for the following analysis and evaluation of hot spots of transparency issues in the food chain. This analysis builds on a detailed description of the food chain and a structured description of major transparency issues related to food safety (see Hofstra et al., 2010; project deliverable 3.1 and Knorr et al., 2011; project deliverable 3.5). A number of key thematic areas were identified, namely emerging risks, emerging technologies, food safety governance, and the parallel economy.

Emerging risks: It is well established that intentional and unintentional alterations in practices of primary production such as harvesting, sourcing, preservation, processing, and packaging might have consequences in the production of, or the selection for, new, unforeseen risks. Likewise, new analytical and other scientific capacities can identify, or make more explicit, risks which had previously been unrecognized. It is known that consumers have particularly strong and deeply-felt concerns about chemical contamination with delayed pathological effects. Food contact materials are examples of potential sources of exposure to hazards of this type which are, where recognized as representing a risk, highly controlled. Certain packaging systems in which food contact is accentuated are indeed specifically demanded by supermarkets (e.g., individual, blister and vacuum packs for the cold chain, interleaved sliced hams, and cheeses). The consequences of emerging risks in the food chain vary according to the specifics but there are certain scenarios that are similar for any of the sources of such risks. Certainly, a drop in confidence in the food chain ensuing from any outbreak or incident would occur independent of the specific cause. However, while the level of reduction in consumer confidence and trust will depend on a number of factors, an unforeseen new risk will always have a particular impact.

New technologies: New technologies include, but are not limited to, the physical processing technologies, some of which have been developed and extensively studied over the past few years. In some cases, the technology itself is not novel at all but the application presented is (such as the shelf life extending combinations of microfiltration and pasteurization in milk). Where the application of a novel technology and its communication to the consumer could provide a competitive advantage, its link with food safety issues may lead to considerable transparency challenges. As an option, one may directly mention food safety in communication with consumers, although this is known to be rare at present. It is more common that only some parts of the message might allude to food safety aspects such as the extension of shelf life or the limitation in the use of additives in production. For many novel technologies, elements of potential interest for consumers, such as indicators of the impact on food safety of food processing (process parameters) or data capture, are still incompletely defined. As an example, "milder" food processing treatments tend to exert their antimicrobiological effects in a more subtle and directed manner than the classical (normally thermal) methods they are designed to substitute. In such cases, it is likely that the efficacy of these effects will be sensitive to variability of the food matrix and thus each new application would require careful validation.

A further issue in this domain pertains to the "ownership" of the "brand" of the new technology in generic terms. Where the specific technology in question (e.g., specific antimicrobial edible film) is subject to industrial property protection, the owner of this property will be careful to assure that its application does not prejudice the value of the technology. This is very clearly illustrated in the case of patented, functional ingredients from which functional or health claims can be made. In these cases, the owning company normally controls completely the use of the technology with legally binding and extremely explicit limits to the way the technology is used and presented to the consumer. However, the general acceptance of a new range of specific applications can, of course, be prejudiced by poor performance or, more dramatically, safety issues concerning a specific technology from this generic range. While being obvious, this is a risk which needs to be addressed if new technologies are to be accepted by the consumer.

Food safety governance: Since the publication of the *White Paper on Food Safety* in 2000, all legislation relevant to food safety has been in the form of regulations. This implies the direct incorporation of the EU legislation into the legal systems of the member states. The capacity of individual member states to implement and apply this legislation varies considerably for a number of reasons. The EC supports comparative implementation through the *Food and Veterinary Office (FVO).* This approach is highly transparent—with all relevant reports and replies being published in web form. Variations in the application of legislation will thus tend to diminish over time but some will undeniably remain, including those which are embedded in cultural customs.

The widespread application of HACCP principles (as demanded in the EC Regulation 852) provides an example of how national bodies

(Competent Authorities) are required to "*fill-in*" details which the legislation purposefully leaves out. These differences in specific levels of stringency and detail take on a particular importance when the food in question follows the traditional supply chain and reaches the consumer without passing through one of the modern large-scale producer or supermarket brands (see later). In numerical terms, it is possible to determine the relative capacity of any member state to successfully implement the new (and new style) legislation—the numbers of staff at the various levels in the relevant functions are verifiable. Formalized evaluations of the capacity and actual activity of national competent authorities are already carried out by the FVO, albeit with a remit to assure harmonization through programmed "policing" activities. However, detailed knowledge as to the underlying causes of variation and the effects they might have on intra-EU, cross-border trade is still incomplete. Variation in stringency and rigor in the application of legislation leads to an environment with a diminished competitiveness, especially for the smaller producers who intend to trade within the EU27 space.

When food products pass through modern, brand-driven chains, the levels of stringency are more harmonized across national borders. This is essentially due to the fact that major brand holders will stipulate key specifications for food safety, often anticipating or exceeding legally imposed levels of protection according to current scientific knowledge. In addition, the rigor in the application of the controls imposed is also greatly harmonized thus further reducing country-to-country variations (see, e.g., www.brcglobalstandards.com, www.ifs-certification.com). Commercially-imposed controls are characteristically "nonnegotiable" in the sense that once applied, either by the brand holder or the "owner" of the certification scheme, their compliance is obligatory in order to have access to the chain. Very often the specifications demanded imply extra costs for the suppliers.

Although largely based on risk analysis and on scientific principles, there is no obligation for such approaches to be followed. Customers may formulate food safety demands of whatever kind they want based entirely on their own criteria. Providing that any specification affords a protection which is at least equivalent to one required by law, the adherence to such privately-imposed demands is related to extra costs. The lack of transparency in fixing specifications is therefore a cause of tension between suppliers and customers.

Parallel economy: The parallel economy is characterized by distribution activities outside the classical channels, for example, farmers' markets. The inherent variability in the stringency and rigor of implementing food safety legislation due to cultural diversity and economic differences contributes to considerable differences in the way in which the parallel economy operates in the various member states. There may be variations due to differences in opportunities for food sales via the parallel economy—both by differences in the ability to provide appropriate agricultural produce and by differences in the capacity of subsequent stages for serving parallel economy activities.

Priority Goals and Challenges: Food Safety

Goal 1: Addressing transparency issues related to emerging food safety risks.
Challenge 1: To understand consumers' perception of messages they receive concerning the appearance of new risks related to food and to understand how and to what degree consumers can accept new risks without losing confidence in the chain.

Furthermore, there are variations in the way the parallel economy is perceived by different groups of consumers. One can expect a spectrum of degree of acceptance of foods being produced and distributed without passing through the formal economy. This is likely to be influenced, at least in part, by regional and national factors. In rural and rurally-influenced communities, food which is locally produced and perceived as "natural" or traditional is often perceived as being superior to food purchased in the formal economy. In urbanized areas, the existence of an infrastructure of ethnical restaurants and food services building on food prepared according to traditions from all over the world may add to the risks of food safety in parallel economies.

4.1.2 Goal 1: Addressing Transparency Issues Related to Emerging Food Safety Risks

Emerging food safety risks bring up specific challenges related to transparency. On one hand, it is far from clear how consumers judge the appearance of new risks or where they see the responsibility for their emergence. In those cases in which the scientific and technical community knows of a risk but has not had time to understand, translate, and communicate this to stakeholders of the food chain, this could be perceived as a lack of transparency. On the other hand, one always needs to generate new knowledge on the cumulative and long-term effects associated with, for example, exposure to chemical hazards—including those from food contact materials. Furthermore, the progressive revelation of potential risks via improved analytical capacity may act as a vector of emergence, requiring a continuing research effort to ensure that the implementation of control measures accompany the risks and not just the technical capacities. The same can be said of the multiple causes of emergence of microbiological safety risks. Better pathogen detection and description should be accompanied by a greater, broad-based perception of the risks they might represent.

4.1.2.1 Major Research Challenges

Challenge 1: To understand consumers' perceptions of messages they receive concerning the appearance of new risks related to food and to understand how and to what degree consumers can accept new risks without losing confidence in the chain.

Expected outcome

- Description of consumer perception of emerging risks in the general context of food safety and in particular in how this can modulate their confidence in the food chain.

Challenge 2: To assure that emerging risks are both technically and scientifically identified as early as possible and that all relevant information is communicated to stakeholders in ways that are appropriate to them. Mechanisms should be created to rapidly commission research in an attempt to anticipate potential risks and to catch new risks before their potential consequences are felt.

Expected outcomes

- Foresight and scenario studies with experts and modeling approaches to identify and prioritize potential hot spots of emergence of new risks in the food chain.
- Research commissioning mechanisms for the rapid and appropriate generation of knowledge in response to indications of the emergence of new risks.

4.1.3 Goal 2: To Ensure that Transparency Issues Do not Impede Emerging Technologies from Achieving Their Potential

This goal is mostly centered on those technologies that are claimed to represent some contribution to food safety, by partially, or totally, substituting existing technologies while adding value through cost, efficacy, or quality benefits. Technologies might even represent an entirely new opportunity of reducing risk that creates a new range of possibilities. The long history of successes and failures in the introduction of new technologies illustrates that there are intrinsic and extrinsic factors that influence whether these will be accepted or not. This area has received considerable attention in the literature. However, it is important to know the effects of introducing new technologies on the overall confidence in the food chain. It is a precondition for the successful introduction of new technologies with inherent merits to know how they contribute to food safety in the chain in which they might be integrated. This knowledge is required not only for the stage at which they might be applied, but in the context of a full chain risk assessment. Such science-based performance objectives are key instruments in transparency.

4.1.3.1 Major Research Challenges

Challenge 1: To more fully understand how consumers perceive the use of new technologies in the food chain and how their perception affects their trust in food safety. As perceptions might change over time, one needs improved methods for monitoring perceptions and attitudes over time and the establishment of appropriate competences and capacities.

Expected outcomes

- Description of consumer perceptions of emerging technologies in the general context of food safety and of effects on their confidence in the food chain.
- Robust methods for monitoring perceptions, attitudes, and trust over time.

Challenge 2: An expanded and improved definition of the safety contribution of new technologies. This should not be limited to the stage of the food chain in which the technology intervenes, but also to potential impacts on food safety of the food chain as a whole. The outputs should contribute to the specification of how to quantify a technology's performance toward improvements in food safety in the chain. The incorporation of such results into the governance of food safety in a just and efficient manner will be of particular importance.

Expected outcomes

- Improved descriptions of food safety impacts of key technologies in food chains, including the formulation of models which could allow predictive and integrative approaches.
- Improved communication channels and methods for the rapid and fair employment of performance indicators and their quantification into the governance of the food chain.

4.1.4 Goal 3: Providing a Fair and Functional Governance of Food Safety

The responsibility for the governance of food safety is, in reality, shared between official bodies and the enterprise actors in the chain. Safety-driven brand protection specifications driven by large supermarket chains (e.g., BRC, IFS) are very stringent. Safety demands are often in excess of what is legislated and are always nonnegotiable. This stringency is passed back down the production and supply chain. Also brand-holding producers often practice their own, nonnegotiable schemes with suppliers and copackers. Differences in the criteria applied by commercial operators (and between different operators) and official agencies (and among official agencies across the EU space) is a cause of tension. Measures need to be taken to ensure that criteria are applied in a transparent manner and knowledge must be generated on how to ensure that such measures are being successfully applied.

4.1.4.1 Major Research Challenges

Challenge 1: To understand in detail the differences in the way governance practices are employed by the official agencies responsible for food safety in the EU member states and their component regional structures. These practices need to be compared and contrasted with governance practices employed by modern retailers, branded wholesalers, brand holders who employ contract manufacture, and by modern large-scale restaurant chains. Descriptive knowledge needs to be complemented by an analysis of the underlying factors which influence variation.

Expected outcomes

- Detailed description and comparison of the performance of inspection and inspection systems across the EU27 and how they interact with commercial systems.
- Detailed comparison of state and regional governance schemes other than inspection.
- An understanding of the key factors underlying the differences in governance systems.

Challenge 2: To understand the differences in the stringency of the criteria applied in the application of food safety legislation across the EU. In this case, it is not the practices and capacities themselves which are of interest, but rather the actual realizations of process and product specifications which are deemed to be acceptable or not acceptable.

Expected outcomes

- Detailed description and comparison of acceptance/rejection criteria which are applied (or permitted) across the EU. This will not cover all products or product types but should certainly include a sufficiently wide selection to enable a faithful picture to be drawn.
- Details and analysis of the reasons behind the variation in stringency.

4.1.5 Goal 4: Understanding the Effects of the Parallel Economy on Food Safety

The existence of uncontrolled and unregulated activities in the food chain is a concern for those responsible for food safety assurance. By avoiding official recognition, chain actors can simply ignore many of the demands that ensure that food is consistently delivered in safe conditions to consumers. This applies to all parts of the chain reaching from primary production to food services and involving activities such as packaging, manufacturing, laboratory, and consultancy services. Thus, the parallel economy can impact negatively on food safety in a number of ways.

4.1.5.1 Major Research Challenges

Challenge 1: To understand the role of cultural and regional diversity in the development of the parallel economy in the EU. It is known that different cultures view the parallel economy in different ways. While some are relatively permissive, others strongly reject it. The level of resistance to accepting uncontrolled food does vary likewise. In the interests of transparency it is necessary to understand how food hygiene might be impacted by incorporating food into the chain which is from regions that are culturally more permissive in this respect.

Expected outcomes

- An analysis of the magnitude of incorporation of food into major chains from parallel economy sources across the EU.
- An analysis of the impact on food hygiene resulting from the incorporation of food from parallel economy sources into chains.

Challenge 2: To understand the impact of unregulated labor on food safety. Many hygiene-sensitive jobs, particularly but not exclusively in retail and food service, are unskilled and poorly paid. As some of this employment may be even irregular or illegal, it is even more difficult to assure hygiene knowledge and practices. Particularly in food service, such circumstances can be critical to food safety. Further research is necessary in order to identify the extent of this problem and ways to resolve it.

Expected outcomes

- A definition on the impact of unregulated and high-turnover labor in hygiene-critical jobs within the food chain on food safety.
- Proposals for addressing poor levels of hygiene knowledge among unregulated and high-turnover workers.

4.2 TRANSPARENCY CHALLENGE 2: TRANSPARENCY FOR TRUST IN FOOD QUALITY

4.2.1 Scope and State of the Art

Food quality is a crucial success factor in order to maintain high standard products. An appropriate transparency is a key success factor for the ability of the food chain actors to guarantee a maximum level of food quality. This is due to the complexity of the food chain which may consist of multiple single stages from the production of raw material by agriculture, up to the final distribution by retailers. In addition to the many stages, the food sector is characterized by a great variety of food products and processes as well as by numerous regulations regarding food quality. Furthermore, as food production is not dominated by a few global corporations but builds on a multitude of SMEs, its complexity involves organizational particularities including cultural diversity.

Food quality is a key factor for consumers in their buying decisions. In food, the assurance of certain quality requirements for raw materials and semi-finished goods is the prerequisite for achieving maximum end-product qualities within a multistep production process. The reliability of food quality controls depends partly on time-consuming and cost-intensive methods and procedures which affect their application and use. Hence, recent emphasis was placed on the development of rapid, cost effective, and preferably nondestructive techniques that find increased application in the food sector. Furthermore, although a variety of analytical means is available for the measurement of quality attributes, their correlation to consumer perception remains a challenging task.

The availability and transfer of quality related information within the food chain is directly linked to transparency issues. However, up until now, the transfer of information related to food quality along and within the chain, and the appropriate coordination of this transfer, is still limited. Apart from such deficiencies, the generation and transfer of information related to food quality in industry may result in a number of signals that integrate available information and provide a certain message to recipients. The major food-quality-related information can be summarized in some categories such as chemical and nutritional product composition, sensorial and physiological characteristics, characteristics of the production process, the status of raw materials, contaminants, microbiological quality, and food packaging. However, new complex product formulations are on the market and require the adaptation of characteristics attributed to the aforementioned categories as well as the modification of the available analytical methods for their control.

In addition to new product formulations, new processing technologies are developed in order to overcome some of the disadvantages of thermal food processing such as losses of certain nutrients, formation of toxic compounds or of compounds with negative effects on flavor perception, texture, or color. The application of nonthermal technologies for food preservation such as isostatic high pressure, pulsed electric fields, pulsed light, or cold plasma is seen to have the potential to replace or complement traditional food processing by reducing the negative impacts of thermal food processing. Industrial applications in the food industry do exist already. However, these applications face challenges in ensuring the safety and quality of food while ensuring consumer acceptability. Consumers' perception of (mainly industrial) food processing is rather negative, probably due to the large attention placed on the formation of undesired compounds and insufficient knowledge regarding novel processing technologies.

Priority Goals and Challenges: Food Quality

Goal 2: Traditional and emerging technologies: A synchronized assessment.
Challenge 1: Evaluation using process-independent performance criteria based on quality and functionality retention and improvement.
Challenge 2: Translation of consumer perception into manageable industrial scale technologies and innovative products.

A number of research needs derive from the aforementioned aspects in order to overcome the currently existing limitations.

Priority is given to five aspects, discussed in the following section: food chain communication and integration, reevaluation of traditional technologies, development of a synchronized assessment for emerging food processing technologies, improvement of suitable analytical methods, and development of concepts for the update of quality standards and legal provisions.

4.2.2 Goal 1: Food Chain—Better Integration from Farm to Fork

For many consumers, production processes of food may seem to lack transparency. Information concerning product streams (e.g., origin of raw materials, transport routes), quality characteristics or the kind of processing used are difficult to access in most cases. On the other hand, transparency and the availability of this information can be a prerequisite for gaining trust in the food chain.

4.2.2.1 Major Research Challenges

Challenge 1: Managing the higher degree of complexity for multi-ingredient products is a challenge that involves the traceability of complex product streams that may extend over the whole globe, as well as the monitoring of quality parameters. New technologies need to be developed that allow an automated, cross-stage, and gap-less monitoring as well as easy transmission of relevant data. The monitoring of quality parameters in complex products usually needs to build on information related to issues such as the origin, processing, and quality of products that are provided for certain ingredients only. Concepts are required for providing additional information considering a full chain approach to allow a comprehensive view on the food chain and to realize transparency within a complex network of food production. Sensor technology and tracking devices are considered as key solutions in order to make quality related information available and transferable within the food chain.

Expected outcomes

- Analysis of complex product streams and multi-ingredient products. Description of monitoring and transmission systems for quality related data.
- Evaluation of data collection systems such as RFID and development of implementation concepts.

Challenge 2: Optimizing the interaction between all members of the chain and selecting the data needed to create or maintain trust and transparency. Market and consumer research programs are of crucial relevance in this context and for evaluating the relevance of information to be communicated. There is no doubt that important quality information is still available inside the chain, but normally most of it is not visible for the consumer at the end. Its utilization requires the specification and organization of a suitable selection process.

Expected outcome

- Specification of data selection processes and communication strategies for the identification and evaluation of suitable communication processes along the chain.

Challenge 3: Consideration of postshopping consumer behavior for maintaining quality as a basis for transparency on food quality and quality development. Food handling at the point of sale and at the point of use by consumers has a major impact on product quality. Knowledge of consumers' behavior on postshopping product handling (taking into account the different levels of education and different levels of information requests) is a precondition for transparency regarding expected food quality at consumption time.

Expected outcomes

- Description of postshopping impact on food quality considering consumer behavior at the point of use.
- Development of communication concepts toward consumers for improving predictability of postshopping food quality.

Challenge 4: Realizing interdisciplinary cooperation between the different stakeholders, organizations, and research institutes dealing with transparency in food quality. Due to the complex and multidisciplinary characteristics of the food processing sector, cooperation is a major prerequisite for a continuous improvement in the transparency of all processes reaching from farms to the end of the chain.

Expected outcome

- Identification of interdisciplinary approaches required for food quality transparency research and development of adequate communication schemes for overcoming collaboration barriers.

4.2.3 Goal 2: Traditional and Emerging Technologies: A Synchronized Assessment

Traditional food processing and traditional food processes have been widely used in Europe in the past and do still include local particularities. However, in order to respond to the consumer demand for more natural, healthier, and sustainable food products and processes, emerging technologies have been initiated. Reservations regarding food processing technologies need to be overcome by providing information on process performance, risks, and benefits that may support consumer acceptance and trust.

4.2.3.1 Major Research Challenges

Challenge 1: Reevaluation of traditional food processing. Reinventing these processes requires the understanding of the traditional process mechanisms and subsequently their transfer and upgrading to modern industrial processes. A reevaluation of existing technologies from a food quality point of view seems essential. Novel processes have to undergo an intensive evaluation regarding toxicological risks. There is no systematic approach for the existing traditional foods and certain critical points, e.g., the formation of acrylamide, are only revealed accidently.

Expected outcomes

- Analysis of traditional food processing with state-of-the-art criteria applied for novel processes.
- Development of process independent performance criteria based on quality and functionality retention and improvement.

Challenge 2: Integrative food process optimization. An effective integrated modeling of food chains and enterprise units is required for generating and validating information regarding changes in food quality during food production, storage, retailing, and point of use. Issues such as packaging technology play a crucial role for quality changes in logistic, for freshness, and for food safety. Therefore, product–process interactions and their impact on product quality as well as the role of packaging and quality changes during storage, retail, and final consumption remain core tasks that require future research.

Expected outcomes

- Setting up of data generation and validation concepts for food quality evaluation during food production and storage.
- Quantification of product–process and product–packaging interactions considering impact of logistic concepts as well as point-of-sale conditions.

Challenge 3: The establishment of a synchronized process assessment scheme including the development of criteria for the analysis and evaluation of process performance of emerging technologies. A reevaluation of product specifications and analytical means may be required in order to adapt them to the requirements of novel products. The lack of information on inactivation kinetics and reaction mechanisms of nutrients, toxins, allergens, microbes, and viruses, shelf-life studies, epidemiological studies, effects on digestibility, on allergens, phytochemicals, and melanoidins clearly indicates further research needs regarding emerging food processing technologies but also regarding traditional food processing.

Expected outcomes

- Process performance criteria and indicators with regard to novel technologies.

- Development of standard protocols for food processing by novel technologies considering industrial as well as research conditions.
- Reevaluation and validation of product specifications and available analytical tools.

Challenge 4: The development of manageable industrial scale technologies for translating consumer perceptions into innovative products is a key step for the further successful development and integration of emerging technologies. Scientists and technologists, but also policy makers with both converging and diverging views, need to become more aware of societal perspectives and both understand and address these throughout their work.

Expected outcome

- Establishment of communication concepts in order to improve consumer knowledge and confidence in novel technologies.

4.2.4 Goal 3: Analytical Methods: Improving Speed, Detection Limits, and Process Adaptation

The improvement of analytical methods contributes to an improved availability of information and may increase the level of transparency. Fast and nondestructive methods for quality analysis need to be further developed as a basis for immediate quality control and management.

4.2.4.1 Major Research Challenges

Challenge 1: With regard to emerging technologies, a reevaluation of current analytical means will be necessary in order to prove their suitability to characterize relevant process–product interactions. New analytical and other scientific capacities can identify risks which had previously been unknown.

Expected outcome

- Characterization, evaluation, and development of analytical methods regarding their ability to identify process–product interactions during processing using emerging technologies.

Challenge 2: Fast and nondestructive methods with appropriate detection limits. Speed enhancements in terms of sample throughput and analytical time requirements are necessary in order to increase the total amount of samples tested and to improve the response time. This would allow both a higher degree of reliability and the ability to undertake immediate actions in case of quality or safety issues.

Expected outcome

- Development and implementation of nondestructive tools for food quality control along the different steps of the food chain.

4.2.5 Goal 4: Improving Food Quality Standards and Making Provisions more Stringent

In times of globalization and international product streams, long transport distances and a complex traceability problem contrast with increased consumer need for high level of quality, safety, and transparency. Improvements in quality standards and legal provisions are necessary for increasing and strengthening consumers' trust.

4.2.5.1 Major Research Challenges

Challenge 1: Advances in scientific knowledge and the development of new analytical methods are the basis for improvements in food quality and safety and should be implemented into the requirements of quality and safety standards without delay in order to guarantee optimal food quality at any time.

Expected outcome

- Concepts for regular updates of quality and safety standards taking into account continuous technological as well as analytical developments.

Challenge 2: Clear, unambiguous provisions in labeling. Surveillance reports show that major deficiencies regarding requirements and consumer complaints are related to labeling. Although a step in the right direction was made by adopting the new European food information regulation, much more stringent legal provisions are needed for preventing loopholes and for improving the reliability of food labeling. The transformation of available product and process related quality information into signals related to consumer information needs remains a core task. In addition, nutritional food quality is strongly related to food composition including aspects such as the energy density or the content of bioactive ingredients. The choice of food suitable for supporting a healthy diet is a daily challenge for consumers to be performed based on their knowledge and on the basis of information on food composition. Hence, the labeling of ingredients, food composition, and nutritional parameters is required in a clear and transparent way in order to allow consumers to evaluate the healthiness and suitability of any food for their diet.

Expected outcomes

- Concepts for the transformation of quality related information into food labeling oriented toward the needs of different consumer groups.
- Development of indicators in order to allow the evaluation of the healthiness and suitability of a given food for individual diets.

4.3 TRANSPARENCY CHALLENGE 3: TRANSPARENCY FOR TRUST IN FOOD CHAIN INTEGRITY

4.3.1 Scope and State of the Art

Ethical, social, and environmental impacts are important for building trust in the food chain, yet they cannot be measured on the food product as such. Thus, the *integrity* of the food chain relating to these aspects must build on transparency (Figure 4.2).

The minimization of negative impacts and the enhancing of positive impacts of social, ethical, and environmental aspects of food chains are increasingly becoming important values around which food choices are made. Communication of these values relies to a great extent on processes of *transparency*. These processes are varied but can rely on tracking and tracing in combination with the use of clear, simple, and up to date information communicated in an effective way. The

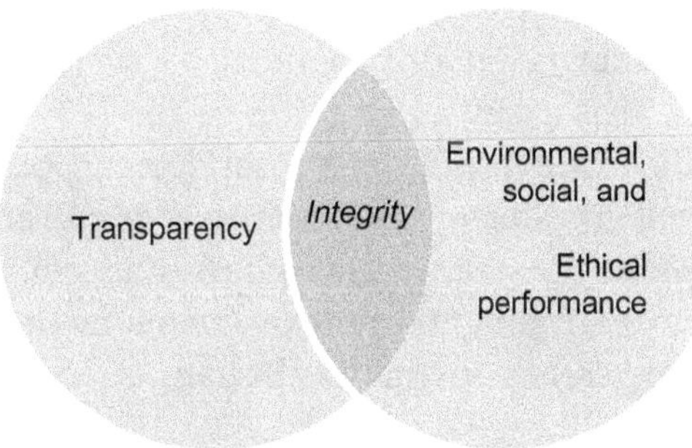

Figure 4.2 Integrity in relation to transparency and social, ethical, and environmental performance.

following builds on an analysis of the state of the art on information use in food chains with relevance for environmental concerns (Öestergren et al., 2010; project report D4.1) as well as for ethical and social concerns (Barling et al., 2010; project report D4.2), and on an analysis, evaluation, and documentation of selected "*best practice*" monitoring and reporting schemes (Östergren et al., 2011; project report D4.3).

On a **company basis** the transparency of environmental, ethical, and social aspects is addressed in two ways. First, by *business-to-consumer communication by labeling* food that is supposed to have certain integrity characteristics, like carbon footprint or fair trade. Second, by *business-to-business information* that ensures that certain standards have been used in producing the goods used in the further processing. One example is *GlobalGAP* that ensures that food is produced on the farm by using state-of-the-art *Good Agricultural Practices* aiming at reducing detrimental environmental impacts of farming operations, reducing the use of chemical inputs and ensuring a responsible approach to worker's health and safety as well as animal welfare.

At the **policy scale** EU has implemented the integrated product policy (IPP) which seeks to minimize environmental impact from products by looking at all phases of a life cycle and taking action where it is most effective. To achieve this objective, the EU IPP is contributing to addressing the environmental challenges identified in both the Sustainable Development Strategy and the Sixth Environment Action Program. The IPP principles have been taken up and carried over by the *Sustainable Consumption and Production and Sustainable Industrial Policy (SCP/SIP) Action Plan* which in turn constitutes a major input to the 10-year framework of the UN/United Nations Environment Program (UNEP) programs on sustainable production and consumption. It is a key assumption of the IPP that the environmental performance of a product or a service can be a factor giving companies or their products a competitive edge, and thus it is a separate aim of the IPP to create the right framework for market conditions that favor environmental improvements in the product chain. An increased transparency in the food chain is crucial to reach this goal.

Priority Goals and Challenges: Chain Integrity

Goal **1:** Valid indicators for estimating the integrity performance within an operational and sound traceability reference unit.
Challenge 1: Establishment of a sound, manageable, and robust framework that lists the relevant aspects to take into account when choosing the traceability reference unit in different types of food chains and covering different integrity dimensions in order to harmonize indicator calculations.
Challenge 2: Methodology to describe different integrity dimensions with appropriate indicators needs to be developed. The fact that different dimensions of integrity do not have the same relevance for all food products needs to be taken into account (e.g., animal welfare).

Another important policy initiative is *the Life Cycle Initiative* which was launched by UNEP and Society for Environmental Toxicology and Chemistry (SETAC). This international life cycle partnership has identified the need for guidelines for social life cycle assessment of products to complement environmental life cycle assessment and life cycle costing, and by doing so contributing to the full assessment of goods and services within the context of sustainable development.

In December 2010 the European Council invited the Commission to develop a common methodology on the quantitative assessment of environmental impacts of products, throughout their life cycle, in order to support the assessment and labeling of products. A communication on this methodology should be adopted in 2012, as part of the revision of the SCP/SIP Action Plan.

Furthermore, the Commission is currently undertaking a study to explore the feasibility of establishing reliable EU Ecolabel criteria for food and feed products. In parallel, the European food supply chain has gathered around the *European Food Sustainable Consumption and Production Round Table (RT)*, an initiative with the objective to establish the food chain as a major contributor toward sustainable consumption and production in Europe by developing a harmonized framework methodology for the voluntary environmental assessment and communication of environmental information along the food chain, including to consumers. In August 2011, the RT carried out a scientific workshop, hosted by the EU Joint Research Center, to discuss methodological issues and recommendations. In their draft conclusions, the RT calls for targeted research efforts to better understand consumer perception, create understanding and action on environmental product information, and for the development of specific guidance on communicating the environmental performance of products.

Hence, while important efforts are taken by private and public bodies to facilitate that all stakeholders in a chain can make informed decisions on environmental, social, and ethical aspects, *based on a life cycle perspective*, a number of issues need further efforts to ensure that all stakeholders can be supplied with the relevant information, and that the framework used allows companies to compete on equal basis.

Furthermore, for the long-term impact and trust it is imperative that there is coherence between what stakeholders perceive as "covered" by a claim, and how the food chain actually impacts on social, ethical, and environmental dimensions. There are, however, huge differences in how comprehensively, accurately, and precisely the different dimensions can be assessed.

Based on an extensive review of state of the art, considering the integrity of the food chain with respect to selected major social, ethical, and environmental dimensions (Table 4.1) and their transparency along the food chain, three major goals to be achieved through focused and intensified research efforts have been identified:

1. Valid indicators for estimating the integrity performance within an operational and sound traceability reference unit.
2. Cost-effective systems for data collection and sharing that take advantage of existing data collected through a food chain.

Table 4.1 Selection of Major Social, Ethical, and Environmental Dimensions of Integrity
• Animal welfare (welfare of livestock and wild animal catch from rearing/capture up to and including slaughter) • Methods of production and processing (organic, IPM/IFM Halal, GM, and Kosher) • Environmental and ecosystem impacts (sustainable agriculture codes, IPM/IFM; fisheries stewardship; natural resource protection; carbon foot printing/labeling; water stewardship, etc.) • Terms of trade (fair price for producers and suppliers, fair trade, fair contract terms, etc.) • Working conditions (labor standards, worker safety and working conditions, hours of work and wage levels, etc) • Social capital (utilization and building of social capital of farmers and growers and of communities)

3. Robust concepts for guaranteeing the integrity performance of different food chains.

4.3.2 Goal 1: Valid Indicators for Estimating the Integrity Performance Within an Operational and Sound Traceability Reference Unit

Assessment of the performance of environmental, ethical, and social dimensions such as environmental impact or animal welfare will often be performed on a farm over a long time interval. In a transparency perspective, this might not in itself be interesting, as a consumer or a company buys one piece or a smaller quantity of goods. Thus, information regarding the integrity performance must follow the relevant quantity, which again must be traceable along the chain, carrying information from the farm to processing to retail. A *traceability reference unit* is the quantity of product for which a specific integrity performance assessment or claim is valid being in the form of either a specific assessment (e.g., a carbon footprint calculation) or a label based on management practices (e.g., animal welfare or organic).

The traceability reference unit thus refers to what level of aggregation is relevant; is it one single pig, all pigs from one batch, or a regional average pig during one year? Moreover, whatever the choice made, data are connected to different batches of products, but batches may be amalgamated later in the process. For example, environmental impact data are often calculated over a year on a farm and are presented as an average, from a number of farms or a country or a region and sometimes also as an average of several years. This aggregated information can rarely be used to distinguish between similar products and therefore does not allow for a benchmarking process, which would help in using the best performing cases as models for others. Finally, the different integrity dimensions require a certain scale in time and space in order to make meaningful assessments.

It is neither trivial to set the relevant traceability unit in food systems nor to quantify the integrity dimensions in question, and these two aspects are related because of the data needed. The formation of an indicator to be used as a signal of integrity and the selected information behind this is not always based on well-established and transparent methods but rather on normative choices or strategic selections and assessments made by the owner of a product or brand.

A large number of tools have been developed to assess and express environmental performance using a variety of indicators and—to a lesser extent—animal welfare. Some of these may be used for benchmarking among farms, for example, nutrient use efficiency or greenhouse gas emission per kilogram of milk. But little research has been done to develop such benchmarking tools on the product chain level.

It is important that definitions and decisions on "relevant traceability reference unit" are scientifically sound and clearly structured. Product category rules (PCR) (see, e.g., www.environdec.com) could be a starting point to define relevant traceability units. There are systems implemented where the process to develop a PCR is described, but because a PCR could become a document with a great impact it is extremely important that the process and resulting document is of highest quality and based on sound scientific knowledge.

4.3.2.1 Major Research Challenges

Challenge 1: Establishment of a sound, manageable, and robust framework that points out the relevant aspects to take into account when choosing the traceability reference unit in different types of food chains, and covering different integrity dimensions in order to harmonize indicator calculations.

Expected outcomes

- Descriptions of how traceability reference units should be defined and the processes needed to reach decisions for different purposes, e.g., *business-to-business information* versus *business-to-consumer information.*
- Assessments of how the choice of traceability reference unit impacts the evaluation of integrity performance and options for benchmarking.

Challenge 2: Methodology to describe different integrity dimensions with appropriate indicators needs to be developed. The fact that different dimensions of integrity do not have the same relevance for all food products needs to be taken into account (e.g., animal welfare).

Expected outcomes

- A compilation or "state-of-the-art matrix" of existing knowledge on indicators for different dimensions of integrity for different types of foods and the key inventory parameters determining the integrity performance of these.

- Animal welfare indicators established that reflects the welfare of the animals used in the production and at the same time are transferable through the chain, linked with an appropriate traceability unit and communicable to consumers and other nonexpert stakeholders.
- Indicators for working conditions (in a broad sense) in the same way as for animal welfare indicators above.
- New comprehensive indicators where present indicators do not reflect the impact aimed for in a sufficient and comprehensive way.
- Recommendations for how the identified indicators could be incorporated into structured processes and tools for quantifications (e.g., PCR).

4.3.3 Goal 2: Cost-Effective Systems for Data Collection and Sharing that Take Advantage of Existing Data Collected Through a Food Chain

In performance-based schemes, like the carbon footprint, the information related to a product needs to be present at each point of the supply chain in order to accomplish a final assessment of the product presented to the consumer. In principle, the information (values and principles for calculation) can be stored and made available to all within the chain and outside the chain. However, some parameters may contain protected knowledge, which the company may not want to share, and some data may have little interest to end users.

Environmental impact data sets are generally aggregated from a very large number of data points supported with a limited number of background data. For transparency, a reported value, associated with a product, needs to be supported by a large amount of data and metadata according to existing standards for LCA (methodology) and EcoSpold (data format). Further on, aggregation of information and data may be necessary in order to communicate information to consumers and nonexpert stakeholders in an understandable way.

For rule-based systems, other types of data and information are generated, for example, through an inspection process. For this type of nonquantitative information, no appropriate system seems to be developed that effectively can be used to communicate such information.

To promote the development of information exchange, *open access databases* need to be established and maintained in order to fulfill the need for open, robust and reliable data/information.

4.3.3.1 Major Research Challenges

Challenge 1: Establishment of a framework for cost-effective systems for data sharing that allows connection to a relevant traceability reference unit and allows a timely and transparent update of process information and a reasonable degree of open access for all interested parties.

Expected outcome

- Standards and guidelines for data formats, data sharing, and data verification that match the traceability reference units developed earlier, take advantage of already existing structures for data exchange within the food chain, do not compromise commercially sensitive information, yet allows for high-quality environmental assessments.

Challenge 2: Identification of barriers and opportunities for making inspection results from rule-based systems publicly available in a meaningful way.

Expected outcome

- Inspection procedures enabling "third party access" to relevant results from inspection and verification processes in rule-based systems.

4.3.4 Goal 3: Robust Concepts for Guaranteeing the Integrity Performance of Different Food Chains

The coherence between what stakeholders perceive as covered by a claim and how the food chain actually impacts on the integrity dimension in question must be assured.

Most integrity dimensions are seeking to address more long-term and societal concerns and are less linked or allocated to the very specific batch of a product which is more important when considering, for example, food safety. A number of integrity aspects *may* not be important to link to a specific small quantity of product from a consumer perspective, but can be assured on a more general level (e.g., animal welfare on a system level) which presently is the case for a number of labeling schemes.

Furthermore, from a consumer perspective, transparency is about creating trust (and the task of the monitoring system is to make sure that this trust is justified). Thus, fully open systems *may* not be needed. "*Transparency on demand*" may be a solution where information is retrieved from a third party (e.g., a database containing confidential company-specific data) and processed on demand to a format that is acceptable and understandable for the customer without revealing sensitive business information.

A valuable way to create transparency, as an alternative to a routinely quantification of often-relative narrow indicators, to estimate the integrity impacts, may be an independent assessment of the performance of farms, processors, and other actors in a food chain through a label scheme. However, this requires a robust and preferably standardized method for assessment of results and suitable method for comparing labels in terms of their performance on important integrity aspects (benchmarking of labeling schemes). A first attempt at such a standardization is done under the concept and organization of International Social and Environmental Accreditation and Labelling (ISEAL) Alliance. However, more research is needed for this in terms of solid basis for indicator choices, aggregations, data requirements, etc.

4.3.4.1 Major Research Challenges

Challenge 1: In-depth understanding on how existing schemes and their rules and practices translates into true impacts that can be communicated to the consumer and to provide feedback to the owners of the schemes.

Expected outcomes

- Independent scientific studies mapping the performance of existing schemes.
- A knowledge base for creating state of the art science-based schemes for increased trust in the food chain.
- New comprehensive indicators for comparing and communicating the long-term impact of different schemes.

Challenge 2: Establishment of criteria to be used in guidelines for external reviews and assessment of schemes in order to facilitate comparability between schemes and over time.

Expected outcomes

- Robust concepts for external reviews of the integrity performance of different food chains.
- A critical checklist detailing the considerations and key questions that external reviewers may need to ask when attempting to compare schemes.

Challenge 3: To develop a framework for information management along the food chain for increased integrity, trust, and business opportunities.

Expected outcomes

- Methodology for mapping consumer trust in relation to content and communication of integrity indicators.
- Recommendations on how to manage the communication of environmental, social, and ethical information to create consumer trust and promote business opportunities.

4.4 TRANSPARENCY CHALLENGE 4: SIGNALING INFORMATION TO BUILD CONFIDENCE AND TRUST IN THE FOOD CHAIN

4.4.1 Scope and State of the Art

The objective of this challenge is to identify information that relates to environmental, ethical, and social impacts of actions and processes in the food chain, to determine the potential, the deficiencies, and the research needs. This will allow the food chain to transmit information related to such impacts toward consumers and policy.

Greater and more appropriate forms of transparency in food chains are a potential facilitator for innovation and change to more sustainable food chains and a more sustainable food system. That is a food system that is more sustainable environmentally, socially, and ethically, and, ultimately, economically. Information in food chains and the transmission of that information in ways that are effective through being informative and understandable to the recipients can result in public-policy-desired behavior change by the food chain and by consumers. Measurable improvements in sustainability impacts are facilitated by improved and suitable transparency.

Food chains will need to move toward transparency that promotes the disclosure of relevant and usable information from food chains to the wider public. Such transparency will provide more symmetry in information flows and allow the sustainability metrics and methods employed to improve over time as the information becomes more relevant to better environmental and social outcomes. That is, the information signaled will result in greater public understanding of the sustainability attributes of food products and so facilitate informed choice by consumers of more sustainable food. The signals deployed to consumers are largely in the forms of product labels and logos (such as based on certification schemes) as well as in-store information and information campaigns, and business advice call lines for customers.

The emergence of food integrity issues that directly contribute to more sustainable food have become of importance to societal interest groups and to groups of consumers, and so to the key buyers in food chains such as the large manufacturers and retailers. This has seen the rise in private standards-based certification schemes with accompanying logos going beyond food safety to include: fair trade foods, the carbon footprint of food products, catch from sustainable fisheries, salads and crops from integrated pest management schemes that enhance biodiversity, animal welfare-based production schemes, and so on. The emergence of these certification schemes has led to the generation of more detailed information of food products.

The signaling of such relevant information to the public and consumers involves a complex set of processes of transmission. Our understanding needs to go beyond the simple and predominant business-to-business (B2B) and business-to-consumer (B2C) models. Equally important are the transmissions of information from business to business and then on to the consumer (B2B2C). Also, rapidly emerging is the importance of social networks in relying consumer-to-consumer information and opinion (C2C). Hence, a more realistic transmission sequence of information that is recurring, and that transparency has to enlighten, is the B2B2C and C2C transmission of information.

Furthermore, there are other social and professional intermediaries who interact with the public and impact upon the information flow to consumers. Civil Society Organizations, such as NGOs, seek to influence consumers' knowledge and decisions through information campaigns, usually from a particular value perspective (e.g., animal welfare). Professional groups such as veterinarians, nutritionists or dieticians may offer information either as independent professional bodies and networks usually not from within the food chain or as employees of the food industry.

A deeper understanding of these processes of information transmission and of the role of food chain stakeholders and their perceptions around information transmission, such as the role of food integrity-based certification schemes and their logos and the ways and rationales for their uptake by key actors in food chains (e.g., retailers, manufacturers, and food caterers) are needed.

Priority Goals and Challenges: Signals for Communication

Goal 1: A more sustainable food chain that utilizes transparency in signaling its sustainability criteria from business to business and on to the consumer.
Challenge 1: To adopt efficient methods of communicating sustainable food choices from the food chain to the consuming public through the effective promotion of appropriate information transmission and signals from the food chain.

How consumers understand and interpret the information that they receive, and the methods they use, and the cues they seek for interpreting and managing such information is vital to realizing informed choice in the food market place. A better understanding is needed of how and to what extent consumers, and specific categories of consumers, understand and interpret the signals around food integrity and the attached environmental, social, and ethical information. This understanding will allow the food chain to find ways to respond in providing the means for more informed choice on these issues. In addition, new forms of information technology are making information from the food chain more accessible and available to the public and in particular to the consumer, and for consumers to exchange information and views among themselves. Future applications of such technologies in forms accessible and usable by consumers and the public will enhance transparency and the goals of a more sustainable food supply. The state of the art of our understanding in relation to key research challenges is elaborated further in the next section.

4.4.2 Goal 1: A More Sustainable Food Chain that Utilizes Transparency in Signaling its Sustainability Criteria from Business to Business and on to the Consumer

Key actors in food chains manage and edit the information about food products offered for sale to consumers. Within contemporary food chains, not least in Europe, retailers and food service companies are key gatekeepers between the consumer and the rest of the food chain. Retailers and some manufacturers have responded to societal and market demands for food integrity around a range of environmental, social, and ethical criteria. In part, the response has been a selective uptake of independent or third-party certification schemes that signal the integrity of the food through logos. These schemes offer retailers and manufacturers a means to construct a narrative of their business profile and brand while conferring third party validation in the logo-based signals sent to consumers.

In order to achieve goals around sustainability, some market innovators (retailers and manufacturers) are seeking to develop their own product supply chains that embed sustainability criteria. Such efforts may be signaled to consumers directly via food products but are communicated also through nonlabel provisions of information, such as annual corporate responsibility and sustainability reports. These "first mover" retailers (notably through the large growth of own label manufactured foods and fresh produce sold) and "first mover" manufacturers are seeking primarily to strengthen their business model to make it more sustainable and resilient; particularly in terms of both their natural resource impact and the resilience of the supply chain sourcing of food commodities and produce.

A secondary benefit is to create or reinforce a brand identity around sustainability issues, sometimes in a selective manner. For public policy, the corporate innovators who are providing a more sustainable food supply are leading the policy and governance response. The interaction of public authorities at national and EU levels with the industry on sustainable food is an important dynamic as is the role of public authorities in framing and coopting the actions of the industry to realize publically desirable goals. One way in which public authorities are seeking to engender greater sustainability is through the use of market-based information for consumers in the form of labels on food, where the authorities are monitoring the initiatives of private sector certification and labeling.

4.4.2.1 Major Research Challenges

Challenge 1: To adopt efficient methods of communicating sustainable food choices from the food chain to the consuming public through the effective promotion of appropriate information transmission and signals from the food chain.

Expected outcomes

- To identify the criteria by which food chain stakeholders adopt food integrity criteria for the products that they sell to consumers and how decisions over the communication of this information are made by the food chain.
- To achieve a more open transparent and effective transmission of information to consumers to aid their choice of sustainable food.

4.4.3 Goal 2: Providing Signals of the Environmental, Social, and Ethical Aspects of Food that Are Understood by Consumers and Respond to their Needs

This goal covers a wide range of issues that are centered on: "signals" and target areas for information disclosure, content of information disclosure and process as well as associated barriers. Widespread efforts to establish signals generally increase disclosure for greater perceived transparency may not necessarily lead eventually to equally evident results across all contexts. Fostering targeted transparency may be: an equally more effective course of social and business actions, prove sustainable business-wise, foster accountability but also positive public deliberations. Target areas may attract increased attention and become hotly debated, likely grounds for uneven media attention, and grounds for political confrontation, thus they are pivotal for ensuring perceptions of greater transparency. Lack of performance (and subsequently, lack of public policy performance) in these areas may be perceived to exist when the nature, speed, and detail of information currently communicated through current adopted information disclosure practices are not actually consumer-useful. Pertinent information/signals (label/nonlabel) may be incomplete perceived as filtered/distorted, nonupdated, nontimely, nonaccessible, discriminatory, proprietary, and thus motivated, at inadequate level of (dis)aggregation, noncomparable, or confusing/inconsistent/nonstandardized.

A pertinent issue is also the activation of conditions for disclosure of further information for elements when they are not part of routine or may become subject of perceived increased hazard/risk. Nonroutine controversial issues of a "crisis" or "problem" nature trigger different consumer attempts of self-control and risk reduction strategies, than with "positive" and good-feeling-induced information/signals. Handling by consumers of information/signals overload and their quality rather than quantity, as well as their balance, may matter. At the same time, food industry, experts, and public policy makers face a substantial set of difficulties and barriers in the process of information disclosure; their views sometimes converging, other times diverging.

The information/signal gap may not be easily bridgeable. The problem may not lend itself to measurement and there is lack of consensus on measurement, so that performance may be feasibly improved. Communication may be impractical, being too multifaceted and complex. Consumers may not have the will, capacity, and cognitive tools to handle complex information given the problem. Variability and uncertainty may not be easily acceptable. Chain members do not easily disclose information due to competition fears despite goodwill intentions, underestimation of consumers' interest or fear of being challenged by clients, and consumers, or complexity of available sources for information. Also, challenges exist in identifying ways to expand current traceability systems to include additional information/signals given the technical difficulties for bulk goods as well as complex multi-origin ingredients-based foods. However, market innovators are beginning to address the multilayered sourcing of their complex food products by setting criteria back along the food chains to the various suppliers.

4.4.3.1 Major Research Challenges

Challenge 1: To identify effective ways of making food chains transparent to the growing demands for information disclosure.

Expected outcomes

- To identify the precise focus areas for consumer requests for information disclosure and the origins of these requests as well as the existence of consumer clusters within or across countries as well as information requirements clusters.

- Better identification of dimensions of information to use (what is timely, relevant, comparable, complete, easy to understand, accurate, valid, and reliable) and how these differ due to the divergence in the underlying nature of pertinent issues (ranging from positive/enhancement to negative/controversial) as well as the process of such disclosed information.
- To identify the actors that will be responsible for identification of pertinent information, channels of information communication, as well as authorities or other actors that will undertake such responsibility.
- A deeper understanding of the precise barriers to information disclosure and efforts for transparency and how to handle these barriers, as they will vary across different food commodities.

4.4.4 Goal 3: Establishing Consumer Trust (the Role of the Media) and Managing the Transition to Greater Transparency

Perceived trust is fundamental and it forms the basis for greater or lesser needs for information disclosure. Yet its complex nature and interrelated interfaces have undermined the full understanding of its functioning for social engineering issues. "Trust" interrelated aspects are linked to several aspects. Trust may be determined or influenced by multiple and interwoven individual and sociocultural characteristics of those who exhibit trust or lack thereof. Trust is determined or influenced by the (perceived) characteristics of the content and amount of information received. Public trust in institutions functions differently according to how a particular risk is managed or communicated; sometimes, the perception of risk appears to be a component of trust, other times referred to as a consequence rather than a determinant of trust, in the sense that, if people trust an institution to manage a specific risk, they perceive the communicated risks as smaller or the benefits as larger.

Trust increasing or destroying features attributable to an institution and those responsible for risk assessment, management, and/or communication, benefit(s) perceptions of innovations and actions as well as issues of credibility and motives is also relevant. Perceived industry motives are important here and regulators' vulnerability compared to third party/independent actors. Moreover, consumer questioning on motives may not limit itself to food supply chains alone. Individual citizens and consumers may distrust the motivations of regulatory institutions under conditions where they perceive regulatory activities to be promoting the interests of specific parties rather than public welfare. The role of media in handling and transmission of pertinent information as well as the involvement of third parties that are percieved as being independent is fundamental here. Furthermore, the issue of unveiling illegal trade through the media either linking to the parallel economy or fraudulent actions needs to be examined further.

4.4.4.1 Major Research Challenges

Challenge 1: How to establish and manage consumer trust, taking into account the role of the media in this process, and managing the transition to greater transparency.

Expected outcomes

- Identification of how to establish suitable high consumer trust reference points and the means and methods to manage these reference points subsequently.
- Better understanding on the exact role of media for consumer-targeted information disclosure and portrayal of supply chain transparency efforts.
- A greater understanding on how to manage the transition from an environment in which (it is assumed that) there is lesser information disclosure ("less transparent") to an environment of greater information disclosure ("more transparent").

4.4.5 Goal 4: The Development and Utilization of Technologies to Facilitate the Flows of Information and Transmission of Signals Thus Enabling Better Transparency

Information and communication technology (ICT) plays an important role in data and information generation and gathering, storage, access, and transmission along food supply chains. Increasingly, signals rely on technology at the points of purchase and postpurchase. This includes the use of hand-held devices and greater use of RFID technology or other technologies in the future. In addition, advances in symantic web technology will allow for much more sophisticated information to be collated, disseminated, and delivered to mobile devices at point of purchase (in store or when ordering via the web), and postpurchase in response to specific questions about products purchased. The growth of "infosumerism" will mean interested and attentive publics demanding more information and transparency about their product purchasing.

Technology has the potential to facilitate participatory forms of transparency and disclosure in which actors along the supply chain can make specific information requests, or, in the case of producers, can allow for greater elaboration on the information currently shared. Thus, for example, 'origin' as a signal has the potential to go beyond a reference to a national location or address of manufacture but can also be a point upon which information about producers and their localities are shared. Consumer engagement on the application and use of these technologies, including their accessibility, is integral to their enhancing the greater uptake of more sustainable food.

4.4.5.1 Major Research Challenge

Challenge 1: To adapt and promote the application of new ICTs that enable and facilitate the potential purchase of sustainable food by the public.

Expected outcome

- The application of ICTs to make information on the environmental, social, and ethical characteristics of food products more readily accessible and understandable for the consuming public.

Challenge 2: To unlock hidden information for utilization by consumers. Products are increasingly linked to labels or certificates of any kind (e.g., eco labels, quality labels). Labels are usually representatives of clusters of information regarding controls, process organizations, product compositions, origins, etc. Intelligent IT devices in consumers' hands such as smartphones may link up with the respective label owners through, for example, the Internet cloud for unlocking the hidden information. A consumer with a device including its profile of interests (e.g., on allergens) may be served directly with the relevant information. Hidden information is not restricted to label information but may include information from legislation as well as information on brands or brand owners.

Expected outcomes

- Overview on hidden information available at major consumer products and evaluation of results regarding consumer transparency needs.
- IT tools for utilizing the concept of unlocking hidden information by consumers.

4.5 TRANSPARENCY CHALLENGE 5: TECHNOLOGICAL BASELINE INFRASTRUCTURE FOR TRACKING AND TRACING

4.5.1 Scope and State of the Art

To facilitate management of production chains, IT-supported tracking and tracing and quality assurance systems have been developed and applied also in the food sector. Existing solutions in most cases focus on a certain production chain or a part of a chain. This has led to a number of information exchange islands with barriers and media breaks between systems. In reality, changes in supply chain configurations and interconnections between chains lead to the transformation of the linear structure into a highly dynamic food sector network. Interoperation of different tracking and tracing systems is thus a prerequisite for appropriate food sector transparency. There is currently no solution available that is suited to all stakeholders within the food sector and that satisfies the requirement of being able to track and trace according to different scopes of a chain.

The input of a system providers working group and further research work has led to an overview of the sector structure, of systems already in place and of properties of food products that may have an influence on how certain functionalities can be technically realized. Stakeholders' expectations and legislative regulations that have to be taken into account in drafting a tracking and tracing infrastructure to support transparency have been analyzed. Existing tracking and tracing systems have been dissected as to what kinds of products can be handled with them, what functionalities they provide, and what methods, standards, and technologies they use, to find out, if there is a common ground upon which a backbone solution can settle.

Within the food sector, a broad diversity of enterprise size distribution characteristics across countries and across different stages of the food chain can be found. While the larger enterprises are commonly small in number but contribute a relatively large part to the economic outcome and to the percentage of bound labor force, small enterprises still play a major role in various stages of the food sector, especially in primary production and specialized retail stores. With this regard, the sector differs from other industries like the electronics sector, where there are almost no small enterprises present in the supply chain, or the automotive sector, where there are lots of medium-sized enterprises in the preproduction parts delivery stage of the chain and a small number of very large corporate enterprises doing final assembly. An important challenge in drafting a backbone solution specification will thus be working out how the scalability requirements resulting from the sector structure can be achieved. Methods and technologies used will have to accommodate on the one hand large amounts of smaller data packages, and on the other hand a large number of small stakeholders.

Stakeholders' expectations and needs as well as their motivations to use a tracking and tracing system differ depending on their position in the chain. Having access to more information than just bare tracking and tracing data is a common requirement to all stakeholders.

Special properties of food and processing and production methods have implications on the setup of a tracking and tracing backbone. A number of problems can be avoided by following certain handling and processing best practices. However, especially in primary production and raw material handling a few difficulties remain leading to research needs mentioned in Goal 2.

A basic information set to enable tracking and tracing can easily be agreed upon as its major requirements are given by legislation, especially EU Regulation 178/2002. Further attributes either to the product itself or to other information stored elsewhere may however be necessary to allow for additional controls or safety proofs and for chains to be separated based on scope (e.g., organic/nonorganic, potential nutritive allergen content). The demands upon providing the possibility to allow for different views and for a number of varying analysis methods are diverse. This has the consequence that not only on the data layer but also on the service layer, careful consideration has to be given to the reusability of data in different contexts. An important aspect to consider is the fact that additional information within a system may also be used to market quality products and can thus serve as an incentive to implementation.

Concerning technical implementation of tracking and tracing systems, web technologies have found their way into systems, but the methods used differ. Nevertheless, there are a number of commonalities among systems and generally applicable methods. The central database paradigm is still widespread in comparison to a distributed storage approach. Mapping existing systems' data content into a distributed, networked infrastructure result in research challenges formulated in Goal 1. It is a crucial factor for success to find a set of standards and methods that are up to the task but at the same time simple, clear, and generic enough to be accepted by everybody. Four aspects have been identified that have to be considered on the technical level in building a backbone infrastructure: identification of items, protocols used in communication, syntax, and semantics of data exchanged. For each, several technologies exist to provide the necessary functionality.

Priority Goals and Challenges: Technology and Tracking/Tracing

Goal 1: Making different subdomain level data encodings intemperate.
Challenge 1: Creating a universal food sector domain ontology by networking subdomain models implicitly given in existing standards, vocabularies, and coding systems.

Complex messaging protocols based on SOAP can be used in well-defined, controlled environments but will probably be too difficult to implement on a larger scale. RESTful web services have been proven to be better suited to networks with large numbers of small, anonymous stakeholders and a lack of control. On the syntax level, XML is already widely used in the food and agricultural sector. It is thus well understood and can easily be implemented by most stakeholders. The disadvantage of XML is its inefficiency during data transfer on the wire due to its verbosity. This may result in problems on large-scale tracking and tracing. There are however replacements for XML that are easy to handle and can be converted without much effort, like JSON, so that syntax issues will not be a limiting factor in implementing the backbone. Data items in the basic tracking and tracing data set are semantically well defined. Giving additional data for enhanced transparency with regard to other aspects however requires a flexible and extensible container that calls for concise formalized and machine readable semantics of its content.

4.5.2 Goal 1: Making different Subdomain Level Data Encodings Interoperate

A crucial part in communication across a network is a common understanding of the meaning of data items allowing for correct interpretation within information technology systems. In simple networks, this problem can be tackled by bilateral agreements. For larger-scale communication purposes, data dictionaries and vocabularies are defined, which can be shared among the respective stakeholders. Within a basic tracking and tracing data set this is of importance, especially for encoding product names and categories. Naming and encoding of further attributes is relevant as soon as additional information on origin or processing parameters is demanded.

A number of data dictionaries, thesauri, ontologies, and encoding systems exist in the food and agricultural sector that focus each on certain subdomains. For the description of food, the multilingual thesaurus LanguaL exists. It offers a framework using facetted classification. Each food is described by a set of controlled terms. The classification can thus be used to derive food's nutritional characteristics. In the agricultural area, the thesaurus Agrovoc, managed by the Food and Agricultural Organization of the United Nations (FAO), is a hierarchical scheme of terms being suited to describe agricultural resources. There are a number of additional, implicit vocabularies available like the one laid down for process steps in the GS/1 Electronic Product Code Information Services (EPCIS) standards. Most of them cannot interoperate with other vocabularies at the moment, i.e., there are no automatic linking mechanisms in place. Relationship definitions between concepts and terms are often missing or too flat to be useful for flexible and dynamic information exchange using, for example, reasoning mechanisms to derive new information and generate signals.

Thus increasing the usefulness of a European backbone solution by offering more than just simple tracking and tracing will reach its limits quickly if this problem is not tackled. The FAO currently provides the knowledge organization system (KOS) registry for collecting and referencing different knowledge organization and sharing systems in, the agricultural and food sector, thus providing a basis from which further harmonization and interconnection work could start. Apart from poor interlinking, it was found that in total, there are gaps concerning the naming of intermediate products, feed stuff, or ingredients which are not food. To facilitate information technology supported sector-wide, interoperability between systems and transfer of existing data and information to human readable and understandable signals concerning environmental, ethical, or social issues, more development is needed to link multilingual vocabularies and to fill the gaps to match the needs of a complete transparency framework.

4.5.2.1 Major Research Challenges

Challenge 1: Creating a universal food sector domain ontology by networking subdomain models implicitly given in existing standards, vocabularies, and coding systems. Finding technical methods to automatically map content of information and data packages to alternative representations, data formats, and information models by using this domain model.

Expected outcomes

- Technical analysis of various food sector thesauri, vocabularies, and coding systems.
- System and infrastructure to network distributed vocabularies.
- Interoperation, mapping, and translation tools based on extensible descriptions of semantic meaning of information items.

4.5.3 Goal 2: Feasible Identification of Holdings, Production Sites and Units and Sound Definition of Traceability Reference Units in Primary Production

Identification mechanisms are a prerequisite for tracking and tracing. This applies to both the objects being tracked and traced and the intermediate steps and locations encountered along the chain. Concerning the moving objects in a tracking and tracing system, the traceability reference unit (TRU) is a common concept describing a collection containing several product units with identical properties. In general, TRU status is assigned to uniquely labeled—and therefore identified—fixed size containers. In an IT supported system, the TRU is also the smallest possible information unit in the way that properties and the possibility to track and trace apply to the whole unit. Therefore, the larger the TRU, the less precise tracking and tracing will be.

Currently, a considerable loss of precision occurs in primary production. This is due to the fact that in production within an open system, a static, controllable TRU is missing. Systems in place consider either the whole farm as one big TRU or they save records at field level. The former approach leads to the necessity to handle large units with an accordingly high amount of information related to it that can only be attributed, with difficulty, to certain subunits like production areas or crops. The latter approach has to deal with aspects like splitting, merging, resizing, and ownership changes of fields. Considering the fact that treatments and environmental influences on fields can have effects on food quality in the long term (e.g., heavy metal contamination), keeping a change history is inevitable. Also lacking is a common identification scheme for fields. Such a scheme should reflect the issues mentioned to allow for easy retrieval of change history information.

On the farm level, unique identification is in place, but it serves several purposes, e.g., registration of establishments rearing laying hens, or the registration of farm animals for veterinary purposes. Therefore, a single farm or another food production unit might obtain multiple registration numbers. The format of numbers differs depending on the country. Identification standards common for supply chain management in industry are not used much in agriculture. It is therefore necessary to overcome barriers with regard to interoperability of different identification systems and simplify usage for SMEs.

4.5.3.1 Major Research Challenges

Challenge 1: Developing good practices for handling traceability reference units that have a change history with regard to properties that may influence the product carried within/upon. Creating an identification scheme based upon a reference information model of change history and proposing an appropriate distribution of responsibility for holding and storing information on TRU properties.

Expected outcomes

- Description of changes to traceability reference units of relevance for tracking and tracing.
- Provision of a generalized reference information model.
- Identifier generation rules taking into account the cases defined in the first deliverable in the references.
- Clarification of responsibilities with regard to keeping the history of changeable traceability reference units.
- Methods for information transfer in case of ownership changes.

4.5.4 Goal 3: Supporting Balancing of Demands for Confidentiality Versus Demands of Open Information

Tracking and tracing of food products requires the storage and retrieval of a substantial amount of data. There is a demand for open information by the customer. On the other hand, companies have an interest in protecting intellectual property like recipes and values of production process control variables. Nevertheless, information like that might be required by other stakeholders or in certain cases by relevant governmental authorities in cases of toxic contaminations. To establish a food transparency system, it has to be clarified how much and which data can be made accessible to the different user groups without violating the confidentiality needs of the producers and still ensuring adequate and timely information of others. With an increasing number of stakeholders having individual transparency and confidentiality requirements on a tracking and tracing system, an accompanying access and authorization system's complexity is increasing in a disproportionate manner. It is therefore required to sketch roles, access rights and data flows in respective request–response cycles and to provide generic mechanisms that can be applied on a large scale.

While an access management system provides a framework for enforcing certain restrictions on who may access what, it cannot prevent usage of data for unintended purposes by valid and authorized stakeholders within the system. An important aspect to consider is thus how to make sure that rightfully received data is not abused. Building blocks may be *access history*, providing transparency on who accessed what and appropriate regulations, and when.

4.5.4.1 Major Research Challenges

Challenge 1: Establishing a food sector stakeholder role system that supports network participants' individual information and confidentiality requirements but on the other hand is feasible to be implemented in an economic manner. Providing a distributed infrastructure to support that role system within a tracking and tracing backbone.

Expected outcomes

- Description of stakeholders with regard to their role as information consumers within the food sector and their requirements and needs.
- Identification of valid intellectual property rights and required information to allow for sound transparency.
- Transfer of stakeholder description to user roles and groups.
- Model for access token distribution: building a public key infrastructure offering means of key distribution, key revocation, and group membership assignment.
- Mechanisms for preventing abuse of data.

4.5.5 Goal 4: Sector-Wide Economic and Technical Feasibility of a Baseline Information Infrastructure

Food products generally have a relatively low monetary value per unit and the profit margins are small. Therefore, the economic effort which can be put into an individual product item is limited. A crucial factor for success of an information infrastructure for transparency is thus low implementation cost for production chain stakeholders. While in large enterprises, necessary IT infrastructures exist that can be used to handle provision of tracking and tracing information to other stakeholders, small and in part also medium-sized enterprises face difficulties.

On the level of networking, they commonly can only rely on temporary or unreliable Internet connections. Although broadband Internet connectivity is becoming more and more common in every country in the EU, the percentage of reliable Internet connections is constantly decreasing. This comes from the fact that on the one hand more connectivity options today rely on wireless technology (UMTS, satellite modems, etc.)—especially in rural areas—and on the other hand there is an increasing number of connections without having throughput and availability guaranteed by either technical measures or appropriate service-level agreements (cf. DSL lines with variable bit rates versus various incarnations of ISDN multiplexed lines like E1–E5 with guaranteed bit rates). It is therefore required to provide an operational model for service provision with simple mechanisms to synchronize local (unreliably available) and remote (reliably available) data pools.

A limiting factor for using transparency solutions is time, especially in SMEs. Current methods of documentation and information transfer by manual capture on paper are time consuming and thus in most cases lead to only the legally mandatory information to be recorded. There is however a large potential of facilitation and time saving by combining and leveraging recent information and communication technologies, e.g., mobile phones, together with two-dimensional barcodes and web services to support SMEs in tracking and tracing by automation of information recording and retrieval procedures. Implementing systems like these requires cooperation and sharing of developments among different solution providers on the software side as well as on the hardware side. A demonstration of basic feasibility by prototypes provides an incentive for further initiative and can serve as a subject of analysis, especially with regard to an applicable business model ensuring participants' expectations of fair share of low profit margins per unit are met.

4.5.5.1 Major Research Challenges

Challenge 1: Developing low-cost commodity hardware technology-based solutions to support SMEs in tracking and tracing. Identifying a suitable business model for service provision within such an environment.

Expected outcomes

- Technology for automation and support of tracking and tracing in SMEs. This involves, for example, data synchronization mechanisms on unreliable links, labeling methods, and data capture and retrieval methods.
- A cost–benefit analysis and evaluation of potential business models.

CHAPTER 5

Good Practice Experiences: An Integrated View

5.1 TRANSPARENCY CHALLENGE 6: INTEGRATION

5.1.1 Scope and State of the Art

Transparency is one of the most complex and fuzzy issues that the food sector is facing. It is widely acknowledged that an appropriate transparency is of crucial importance and a critical success factor for (i) sustainable development, (ii) guaranteeing food safety and quality, (iii) providing consumers with information to support their buying behavior, and (iv) identifying a suitable regulatory environment.

Consequently, transparency is one of the most popular concepts within chain management in general and within food chain management in particular. However, researchers, as well as practitioners, often raise the question of whether more transparency is better. To answer this question, one needs to analyze good practice experiences regarding food chain transparency.

Given the economic (e.g., employment, added value), ecological (e.g., food miles), and ethical (e.g., animal welfare, fair trade) importance of the agri-food business, one of the objectives of the transparent food project was to compile a good practice inventory regarding food chain transparency. This inventory analyzed selected good practices in-depth to (i) help making the concept of transparency more understandable, (ii) provide useful examples from different transparency domains (e.g., food safety, food sustainability), (iii) illustrate the difficulties of transparency, (iv) provide good practice experiences that have proven themselves over time to reach transparency in the food chain, (v) provide good practice experiences where the optimal level of transparency can be delivered more effectively with fewer problems and unforeseen complications, (vi) provide useful examples to improve the average performance of existing transparency systems, and (vii) provide useful examples for all stakeholders within the food chain to develop new transparency systems. Hereby, we focused on transparency needs of consumers, industry, and policy toward food safety, food quality, food origin, and food sustainability (environmental, social, and economic issues).

Priority Goals and Challenges: Best Practice

Goal 1: Developing optimal transparency systems.
Challenge 1: To understand the problem the transparency system tries to address and to identify the goals of the transparency systems.
Challenge 2: To understand the differences in stakeholders' interest regarding transparency.
Challenge 3: To identify the optimal level of information to obtain optimal transparency.

Results from the good practice inventory (Gellynck et al., 2011; project deliverable D6.2) indicate that a number of experiences exists that are effective in addressing transparency issues. However, questions remain regarding more specific characteristics of how transparency systems can be best implemented/improved.

Further, there are gaps that are not adequately addressed by practice or where lack of sufficient information leaves some questions unanswered. We refer to these as "research gaps," and pull them together in the form of goals, major research challenges, and deliverables. Research gaps are identified for the following categories:

- developing optimal transparency systems,
- understanding costs and benefits of transparency systems,
- creating multitarget transparency systems,
- identify reference systems for future scenarios.

5.1.2 Goal 1: Developing Optimal Transparency Systems

The European food system is active on domestic markets as well as international markets. In this food system, innovation is taking a leading role as precursor of competiveness, growth, welfare, and well-being. Researchers, as well as practitioners, model the above relationship under perfect competition, whereas perfect information is one of the assumptions of perfect competition. However, within real-life contexts—especially when competition is optimized under (information) constraint—it is more realistic to consider optimal competition rather than perfect competition. Similarly, it is more realistic to consider optimal transparency systems, where the stakeholders have the information that they need to make decision; however, full transparency is not achieved. As transparency systems incorporate multiple stakeholders, an optimal transparency system should consider different interests (e.g., market versus public authorities) regarding transparency.

5.1.2.1 Major Research Challenges

Challenge 1: To understand the problem that the transparency system tries to address and to identify the goals of the transparency systems: For whom do we want to create value? There is no agreement on how to measure the performance of transparency systems, or how to develop an optimal transparency system. Still, performance can generally be defined as the extent to which goals are achieved. Consequently, evaluation of performance of transparency systems (development of optimal transparency systems) remains incomplete or impossible if the achievement of goals is not taken into account. As such, a transparency system is considered to be optimal (high performing), when the goals of the transparency system are achieved. In this context, the existence of clearly defined goals of the transparency system is regarded as the first and most important step in evaluating the performance of transparency system, or on the development of optimal transparency systems. The goals of the transparency systems are directly linked to the problem the transparency system tries to address in a way that the stakeholders of the transparency system select goals that, if achieved, will resolve the problem. As such, both the right problem definition (what the transparency system tries to solve) and the right goal setting is crucially important for optimal transparency systems.

Expected outcomes

- Analysis of the problem definition in transparency systems.
- Identification of the goals of the transparency systems.

Challenge 2: To understand the differences in stakeholders' interest regarding transparency. In addition to commonly shared interests, conflicting interests of stakeholders may also coexist in transparency systems. The interests of stakeholders are said to be conflicting if they could hinder the achievement of other stakeholders' interest. As such, in order to develop an optimal transparency system, the common and conflicting interests of stakeholders should be evaluated, because optimal transparency systems should build on the common interests of stakeholders, while addressing the conflicting interests in the same time.

Expected outcomes

- Analysis of how the different stakeholders' interests are formed.
- Identification of the conflicting interests of the different stakeholders.
- Identification of the common interests of the different stakeholders.

Challenge 3: To identify the optimal level of information to obtain optimal transparency instead of complete transparency (superfluous information).

Expected outcomes

- Identification and theorization (if not case specific) of the optimal level of information.
- Determination and quantification of the difference between the current situation (benchmark) and the optimum.

Challenge 4: To identify how the optimal level of information can be realized. Firstly, one needs to determine the required governance structures that encourage reaching optimal transparency, that effectively and consistently evaluate transparency performance and provide sufficient support and direction through implementation. Secondly, the responsibilities of the different stakeholders need to be defined. Thirdly, the determination of the return on investment of realizing optimal transparency. Fourthly, after identifying the optimal level of transparency, it is important to have a look at not only information quantity but also information quality (reliability, accessibility, etc.). Hereby, the drivers of imbalanced reliability along the food chain play an important role: What causes unreliability? What are the hot spots for reliability (critical control points)?

Expected outcomes

- Identification of the governance structures (e.g. contracts, vertical integration, trust-based relationship) that help to reach optimal transparency.
- Stress test for the governance structures to understand how strong the applied governance structures are, especially how they would function in various crisis situations.
- Analysis of stakeholders' responsibilities in optimal transparency systems.

- Identification of cost of optimal transparency systems.
- Analysis of the quality of information from an optimal transparency system.
- Identification and analysis of drivers of imbalanced reliability along the food chain.

5.1.3 Goal 2: Understanding Cost and Benefits of Transparency Systems

Stakeholders agree that effective chain management and competitiveness requires a good transparency system. Ensuring transparency throughout the food chain can also present challenges and according costs, e.g. the cost of providing information (recording, communication, etc.), the cost of selecting and interpreting relevant information. These costs of building a transparency system have often been cited as a cause of objection.

5.1.3.1 Major Research Challenges

Challenge 1: To identify the costs and benefits for the different transparency domains (e.g., food safety and food sustainability): How are the costs and benefits distributed in the chain? Is there a balanced distribution?

Expected outcome

- Analysis of the distribution of the cost and benefits for the different transparency domains.

Challenge 2: To identify local, national, international (EU), and global, cost and benefits of transparency systems to determine the value of transparency systems and to analyze possible valorization on third markets (e.g., North America, Asia).

Expected outcomes

- Identification of local, national, international (EU), and global, cost and benefits of transparency systems.
- Valorization of EU transparency systems on third markets.

Challenge 3: To identify the determinants of limited transparency (e.g., trust, power, and dependency) and its effect on costs and benefits. For example, lack of communication skills (e.g., not communicating typical failures, weaknesses, recommended behavior in crisis situations) can result in limited transparency systems.

Expected outcome

- Identification of determinants of limited transparency.

Challenge 4: To determine how to create a balanced distribution of costs and benefits: What governance structures are required, with public and market responsibilities? Who should lead/be the initiator?

Expected outcome

- Specification of required governance structures, and public and market responsibilities to create a balanced distribution of costs and benefits.

Challenge 5: To compare internationally (within EU) transparency systems and to identify the effect of different control systems and different (non)coercive systems on the competitive position.

Expected outcomes

- International comparison of transparency systems.
- Identification of the effect of different control systems and different (non)coercive systems on the competitive position.

5.1.4 Goal 3: Creating Multitarget Transparency Systems

Transparency systems can focus on one target (e.g., economic target by price transparency, ecological target by carbon footprint transparency) or more targets (e.g., sustainability: environmental, economic, social concerns). Addressing more than one target does not result automatically in more transparency. On the contrary, multitarget transparency systems can be confusing during communication and difficult to evaluate. Moreover, these transparency systems often lack a clear focus.

5.1.4.1 Major Research Challenges

Challenge 1: To determine how the different targets can be bundled into one denominator. Transparency systems that focus on only one target, e.g., carbon footprint (ecological transparency), can be easily evaluated. Multitarget transparency systems target different aspects which makes it difficult to evaluate the performance. Therefore, it is important to define a denominator which includes the different aspects of the multitarget transparency system. Moreover, it is important to investigate management tools for the different stakeholders and how to deal with conflicting targets.

Expected outcomes

- Analysis of denominators that bundle the different aspects/topics/domains in multitarget transparency systems.
- Analysis of management tools for different stakeholders in multitarget transparency systems.
- Analysis of conflicting targets in multitarget transparency systems and investigation of how to deal with it.

Challenge 2: To determine how one needs to communicate with different stakeholders (from farm to fork) when dealing with multitarget transparency systems. When more than one aspect is targeted in a transparency system, a clear communication process is essential to prevent confusing messages.

Expected outcomes

- Analysis of the communication process in multitarget transparency systems.
- Identification and theorization of the optimal communication process in multitarget systems.

Challenge 3: To extend food safety toward other management practices: Quality signs for transparency. The objective would be to extend food safety, which focused up until now on labeling and accreditation, toward management practices that include risk and productivity/operations management. Hereby, the question will be how this can be realized in the food sector and more specifically how this can be realized by SMEs.

Expected outcome

- Analysis of extending food safety toward other management practices.

5.1.5 Goal 4: Identifying Best Practice Transparency Systems as Reference Systems for Future Scenarios

Future transparency systems will have to deal with future expectations and the opportunities provided by technology in data collection, communication, and use. It will also have to include database services that could complement individual data management in collection but also communicate within the chain and with consumers.

Database services could provide databases where data of general validity have been collected in advance as a basic input for meeting transparency needs of users. While such databases might refer to data relevant for any stage of the chain, including consumers, they would not have to be communicated throughout the chain but be available wherever needed. Information technology might provide support in communication, in dealing with situations where information collection takes place at enterprises within the chain *after the product has left* the enterprise premises, a situation typical for laboratory testing, on a product's path through the value chain (*monitoring*) and in communication with consumers where technologies of the future Internet might provide new opportunities.

5.1.5.1 Major Research Challenges

Challenge 1: Provide chain information reference processes that constitute future best practice cases for transparency in various scenarios. Based on present best practice, such as reference processes, provide enterprises and policy with a guideline on where to move. This could facilitate communication between the chain members and the enterprise investment decisions toward the future.

Expected outcomes

- Specification of transparency needs in selected future scenarios, selected cultural backgrounds, and levels of education considering major product lines.
- Information processes for selected transparency needs and for different levels of technology considering major product lines.

Challenge 2: Provide a roadmap toward future reference processes. Developments in transparency are a dynamic process where investments support a stepwise improvement. This requires the identification of suitable development stages that balance transparency priorities with investment opportunities for various types of chains considering reach (local, global) and major product alternatives.

Expected outcomes

- Specification of roadmaps toward improvements in transparency for various types of chains and product lines.

CHAPTER 6

Enabling Activities and Needs for Action

6.1 TRANSPARENCY CHALLENGE 7: COMMUNICATION WITH STAKEHOLDERS AND MEDIA

6.1.1 Scope and State of the Art

To serve the transparency needs of consumers related to the sometimes complex characteristics of food products and food processes, value focused, simple, clear, and easy-to-understand messages are necessary. An information overload caused by too many details communicated to consumers who are usually not food experts may result in confusion, risk that key messages are overlooked, and may endanger their perception of being properly informed. As a consequence, information for consumers has to be aggregated (in whatever form) and transformed into a simplified message. However, the message has to be linked with background information that includes the details that the message builds on and that might be requested by consumers. This is a well-established approach in literature where it is being referred to as "drill-down" capacity.

The other stakeholders of the food chain such as retailers, industry, service operators, or policy makers may want to build their decisions on signals that communicate much more detail than communicated in consumer messages.

There is a need for a systematic identification of the content, the level of detail, and the format of signals and messages that meet the expectations of the different stakeholders. Furthermore, there is a need for the development of effective communication strategies, which cover the exchange of transparency information between the different stakeholders, the transformation into signals and messages, and the provision of backup information for serving specific requests. In the development of strategies, the utilization of newly-emerging information technologies characterized by functionalities of the future Internet may provide new opportunities in developing appropriate communication schemes.

Priority Goals and Challenges: Communication

Goal 1: Improving the access of stakeholders to transparency information.
Challenge 1: To understand the preferences of different stakeholder groups for different communication channels, communication tools, and communication formats such as languages and audio–visual opportunities considering present and emerging technologies including those of the future Internet. Analyze the effects of alternatives on perception and trust.

6.1.2 Goal 1: Improving the Access of Stakeholders to Transparency Information

Different stakeholders have different preferences for communication tools for collecting and communicating transparency information. Communication channels are means to ensure that the required information is available for the target audience in the right time, in the right place, and at an affordable price. Communication tools are needed that allow meeting the different needs to ensure appropriate perception and absorption by the target audience. Furthermore, the selection of communication channels and tools has to consider the effects on the recipients' trust in the verity of messages and supporting signals.

The provision of transparency information may build on label-based and nonlabel-based solutions. Nonlabel-based solutions include web-based applications that utilize different communication dimensions such as audio and visual communication opportunities. Such opportunities are especially of interest in communication with consumers who need to pick up information "on the fly" and, if possible, filtered and focused according to their personal preferences.

6.1.2.1 Major Research Challenges

Challenge 1: To understand the preferences of different stakeholder groups for different communication channels, communication tools, and communication formats such as languages and audio–visual opportunities considering present and emerging technologies including those of the future Internet. Analyze the effects of alternatives on perception and trust.

Expected outcomes

- Mapping preferences of different recipients for receiving transparency information from different communication channels and tools.

- Analysis of effects of different communication opportunities on recipients' trust in the verity of signals and messages.
- Analysis of opportunities provided by emerging technologies including those of the future Internet for improvements in transparency, consumer perception, and consumer trust.

6.1.3 Goal 2: Organizational Specification of Efficient and Balanced Transparency Systems with Fitting Levels of Detail

Organizational specification involves specification of the levels of detail and the consideration of a fair balance of interest between the providers and the users of transparency. Consumers and policy makers need fewer details of transparency information than retailers and food industry. Consumers' and policy maker's needs for details are not a constant, but may change with the development of knowledge, with changes in public interest in specific claims, and in crisis situations. The higher interest of users in information details may conflict with the business interest of information providers. There is a need for methods to evaluate how a fair balance can be achieved. Furthermore there is a limited knowledge available on measuring the impact of different transparency communication methods on efficiency.

6.1.3.1 Major Research Challenges

Challenge 1: To understand the motivators of different consumer groups for requesting transparency information and to develop methods for determination of the optimal level of details of transparency information to consumers, policy makers, and media.

Expected outcomes

- Identification of motivators of different consumer segments for requesting transparency information including European cross-cultural differences.
- Methodology for identification of the optimal level of details of transparency information to different consumer segments, policy makers, and the media and their changes in crisis situations.

Challenge 2: To understand which factors influence the fair balance between transparency needs of the recipients and the needs of the information providers.

Expected outcomes

- Concept for evaluating the validity of the transparency needs of different stakeholders.
- Identification of the necessary level of guarantees for different stakeholders.

Challenge 3: To evaluate the impact of the efficiency of transparency communication.

Expected outcome

- Methods for measuring the effectiveness of transparency communication.

6.1.4 Goal 3: Improving the Exchange of Transparency Information Between Consumers and SMEs

Consumer requests for transparency information are increasing. Enterprises, especially SMEs, may have difficulties in meeting the transparency requests provided by retailers and the society because of lack of knowledge, lack of resources, and lack of facilities.

6.1.4.1 Major Research Challenges

Challenge 1: To improve the capabilities and facilities of SMEs for communication of transparency information.

Expected outcomes

- Concepts for ensuring low-cost facilities for SMEs to provide transparency information.
- Training programs for increasing the knowledge and skills of SMEs for effective provision of appropriate transparency information on the products and activities.
- Advanced best practice guide for provision of transparency information.

Challenge 2: To improve consumers' understanding of the concept and use of transparency.

Expected outcome

- Consumer education programs on accessibility and interpretation of transparency information.

6.1.5 Goal 4: Establishing Open Innovation Exchange Between Consumers and Members of the Chain at Various Stages of the Chain

Open innovation in the food sector describes a concept that builds on direct communication between members of the chain and consumers as the chain's final customers. This direct communication may support enterprises in the identification of strategic innovations and, in turn, in strategic innovation regarding transparency.

6.1.5.1 Major Research Challenges

Challenge 1: Analyzing and evaluating open innovation concepts for suitability regarding innovations in transparency in different scenarios. Identification and experimental evaluation of most suitable concepts for utilization in industry.

Expected outcomes

- Overview of open innovation concepts with suitability for the food sector and evaluation of their performance.
- Identification of most suitable concept based upon analysis and experimental evaluation.

6.2 TRANSPARENCY CHALLENGE 8: DEALING WITH CLAIMS AND DATA OWNERSHIP

6.2.1 Scope and State of the Art

Data and claims are based on ownership. The use of data as well as the use of claims is subject to approval by owners. Furthermore, if used by actors in the chain, the utilization requires some understanding of their reliability.

Product characteristics that cannot be measured at the final product are principally provided as "claims." The reliability of claims is of critical relevance for the evaluation of a product's characteristics and for consumers' trust in the claims. Its specification and control is especially complex in issues that are difficult to measure and quantify. Of specific importance are claims that incorporate clusters of information such as certificates. In transparency developments, the availability and use of claims is of high relevance. They usually provide clusters of information that could be picked up from the claim wherever needed and without a need to communicate the individual information items across the chain. Furthermore, the ownership of claims puts responsibility for substantiation on ownership, relieving chain actors from the burden of providing information guarantees.

Priority Goals and Challenges: Claims and Data Ownership

Goal 1: Substantiation of claims.
Challenge 1: Substantiation of the reliability of commonly-used claims in the food sector and the analysis of costs and reliability benefits of different control and guarantee systems behind the claims.

Data ownership (see also Schiefer, 2010; project report D7.2) is a critical issue in the food sector. As the distribution of data ownership does usually not match the power balance in chains, it is a source of tension and debate. Data could be owned by individual enterprises, by groups of enterprises, or by the public.

Presently, transparency interests have a major view on data potentially available from agriculture. This includes prominent subject areas such as carbon emissions, animal welfare, and use of pesticides. The collection of data is connected with costs and their use with benefits. Furthermore, data provided by actors in the chain could be used by other members of the chain against their interest. This is part of the debate on the provision and use of data between agriculture, industry, and retail.

6.2.2 Goal 1: Substantiation of Claims

A claim is in itself just a statement that might refer to any of the aspects of relevance for the provision of transparency. For its utilization by any actor in the chain, and especially in communication with consumers, it needs substantiation that supports its reliability and, in turn trust, in the statement's content. This is especially true for claims based on major certification schemes such as *GlobalGAP* dealing with agriculture, IFS dealing with suppliers of retail, and others build on sophisticated controls to provide the requested guarantees. However, even in certification schemes, controls vary between schemes, countries, and certification bodies, which is a challenge for using claims in the provision of transparency one can trust in.

6.2.2.1 Major Research Challenges

Challenge 1: Development of a concept on the specification of the reliability of claims. Using claims from whatever sources in the development of transparency systems requires a unified evaluation approach of the control and guarantee system behind the claims. Such an approach is a prerequisite for motivating enterprises and consumers to accept claims as part of a trusted transparency system.

Expected outcome

- Specification of a standardized concept for the evaluation of the reliability of claims.

Challenge 2: Substantiation of the reliability of commonly-used claims in the food sector and the analysis of costs and reliability benefits of different control and guarantee systems behind the claims.

Expected outcomes

- Substantiation of the reliability of commonly-used claims based on the standardized evaluation concept discussed in challenge 1.
- Specification of costs and reliability benefits of the control and guarantee systems behind major claims and identification of best practice for recommendation.

6.2.3 Protecting and Considering Data Ownership

Protection of data ownership is a core requirement of any communication systems. This is more complex than it might sound. It involves issues of costs and benefits, the protection against misuse of data (e.g., use for purposes not agreed upon in provision agreements), the protection against distribution by recipients to third parties not agreed upon or the protection against access to data not authorized by owners. It is apparent from this list that the protection of data ownership and its consideration in transparency systems requires different disciplines to cooperate.

6.2.3.1 Major Research Challenges

Challenge 1: With increasing interest in transparency, the provision of data becomes an important issue in competitiveness. It is necessary to understand the added value of data and to relate them to the costs of collection, transformation, and communication. This could open the way for the development of information markets where the provision of data is evaluated (and priced) in accordance with costs.

Expected outcomes

- Mapping of data sources and data ownership related to transparency needs.
- Mapping of costs and expected benefits of data collection, data communication, and data use for data owners and data users at different stages of the food chain.
- Mapping of additional costs and additional expected benefits of data collection, data communication, and data use that might be due to emerging developments in transparency in future scenarios.

Challenge 2: The power balance in chains is a source of distrust that could be overcome by contract schemes and organizational developments especially linked to agriculture. Similar challenges are being faced by other SMEs in the chain. However, it is especially relevant for agriculture as one of the major providers of information with relevance for transparency. It is envisaged that model arrangements developed for this stage in the chain could be transferred to other groups as well.

Expected outcome

- Analysis and evaluation of organizational structures and contract schemes that might serve as blueprints for integration of SMEs and agriculture into "fair" transparence schemes.

6.3 TRANSPARENCY CHALLENGE 9: COORDINATION AND COOPERATION INITIATIVES

6.3.1 Scope and State of the Art

A prerequisite for making transparency work is the assurance that all enterprises along the chain adhere to the requirements of a suitable chain information process and are able to link up with their suppliers and customers for appropriate information exchange. Even if information processes have been clearly defined, enterprises still need to be coordinated in their efforts and to have the technological, organizational, and intellectual capability as well as the legal and contractual right to collect, process, provide, and communicate the requested information.

The coordination need is a critical issue in a sector dominated by independent SMEs in agriculture and food industry while to a large extent being connected to multinational enterprises in agricultural inputs and retail. Furthermore, coordination is a prerequisite for any sensible chain transparency initiative as chain transparency cannot be assured by any individual enterprise along the chain. Retail cannot deliver without farms, agriculture cannot deliver without industry and retail. This is a situation where groups have called for policy action to break a possible deadlock.

Priority Goals and Challenges: Coordination and Cooperation Initiatives

Goal 3: Designing markets for information and claims.
Challenge 2: Understanding different options for the organization of markets of claims that are based on initiatives by major providers of claims such as providers of certificates dealing with quality, environmental, social, or ethical issue.

Coordination needs to assure that the technological, organizational, and intellectual capabilities of enterprises engaged in a chain transparency system do fit efficiently together. As an example, if an enterprise communicates necessary information in paper format while the recipient expects information in digital form, the system organization of the partners does not fit and the transparency system involves a communication barrier that needs to be eliminated.

This capability may be referred to as an enterprise's "*T-readiness*" (Schiefer, 2010; project deliverable D7.2). It is a concept derived from the concept of "*E-readiness*" where the focus is primarily on technology. "*T-readiness*" integrates "*E-readiness*" and information content. Serving transparency needs of stakeholders toward the end of the food value chain including consumers requires a level of chain or network development where trading partners operate on a similar level of "*T-readiness.*"

6.3.2 Goal 1: Identifying Suitable Organizational Infrastructures for Coordination Support Toward Increasing Transparency in the Sector

For reaching transparency, there is a need for organizational support. Agreements on the utilization of standards for data exchange, the utilization of technology, the organization and management of interface platforms (data backbones), the evaluation of claims and other issues require coordination in a sector with a diversity in stakeholders, a high percentage of SMEs, and no natural focus point that could assure (from the viewpoint of all stakeholders involved) cooperation and fair coordination initiatives.

6.3.2.1 Major Research Challenges

Challenge 1: Identification of major barriers toward increases in transparency and identification of potential blueprints for possible organizational infrastructures in support of transparency. This might involve new private or public institutional initiatives or organizational developments within the sector.

Expected outcomes

- Analysis and evaluation of organizational infrastructures for coordination support in policy, sector, and enterprise networks.
- Identification of blueprints for organizational infrastructures that could be recommended for promotion in support of transparency.

6.3.3 Goal 2: Reaching a Sector Status in Information Availability and Information Handling that Fits Transparency Needs and Efficiency Requirements

Any coordination initiatives in the sector have to build on enterprises, ability to receive, collect, use, and communicate information that fits the transparency needs of the chain as well as the technological, organizational, and intellectual requirements of the chain transparency system (T-readiness). The organization of an appropriate level of "T-readiness" within the sector is beyond the decision competence of individual enterprises and requires a chain or sector view.

6.3.3.1 Major Research Challenges

Challenge 1: Determining the indicators for a suitable analysis of "T-readiness." Analyzing and mapping the actual level of "E-readiness" and "T-readiness" at enterprise and sector level in the sector. Identification of possible layers of feasible enterprise networks with fitting T-readiness.

Expected outcomes

- Provision of a standardized approach for evaluation of T-readiness at enterprise and sector level and establishment of a category scheme for distinguishing different levels (layers) of T-readiness.

- Mapping different layers of fitting T-readiness in the sector (number and size of enterprises at different stages and in different product lines).
- Analysis of the fit (feasibility) of different layers of T-readiness with different scenarios of transparency needs.

Challenge 2: Analyzing needs for investments and initiatives on enterprise and sector level for moving upward in layers, e.g., moving from a lower to a higher level of T-readiness. Analyzing costs and benefits of selected (content specific) transparency systems of different layers and specification of the benefit–cost ratio of moving from lower to higher levels of T-readiness on enterprise and sector level.

Expected outcomes

- Analysis of costs and benefits of selected transparency systems for different layers of T-readiness considering efficiency as well as transparency needs.
- Costs and benefits of moving from lower to higher levels of transparency.
- Identification of investment requirements for meeting transparency needs of future scenarios.

6.3.4 Goal 3: Designing Markets for Information and Claims

With increasing relevance of transparency information and transparency claims for competitiveness of enterprises and chains, they will have to be viewed as products in their own right and with own market relevance. They might be part of a product's value but might also become issues of independent marketing activities. This asks for the development of appropriate market environments and market rules adapted to different scenarios. Early examples are represented by initiatives like "*book and claim*" where claims are completely separated from the products they initially were linked to.

6.3.4.1 Major Research Challenges

Challenge 1: Understanding different options for the organization of information markets that fit the specific needs of SMEs to trade information for use in transparency systems. Analyze, simulate, and evaluate newly-emerging options that utilize state-of-the-art information technology and may serve customers in the chain as well as database service providers that may offer data of more generic nature to third parties outside the customer range of the data provider.

Expected outcomes

- Inventory of different opportunities for organizing information markets in various domains of interest that fit the needs of SMEs.
- Proposals for organizing future information markets that assure fairness in information exchange while optimizing efficiency and reliability.

Challenge 2: Understanding different options for the organization of markets of claims that are based on initiatives by major providers of claims such as providers of certificates dealing with quality, environmental, social, or ethical issues. Analyze, simulate, and evaluate newly-emerging options that utilize state-of-the-art information technology and provide transparency all along the chain and also toward consumers. This incorporates explicitly market opportunities with consumers that might subscribe to receiving transparency on the reliability and background of claims.

Expected outcomes

- Inventory of different opportunities for organizing markets of claims and claims' guarantees in various domains of interest.
- Proposals for organizing future markets of claims that fit major trading scenarios and major product lines while optimizing efficiency and reliability.
- Proposals for emerging trading opportunities for consumers utilizing newly emerging technologies as well as the capabilities of the future Internet with its cloud functionality.

CHAPTER 7

Concluding Remarks

The SRA provides proposals for future research that could support improvements in transparency and facilitate development toward transparency. However to improve in transparency, research results have to lead to action. Action in a network sector with the characteristics of the food sector needs to build on cooperation and coordination. Being able to cooperate in the sector is a critical success factor. It is an aspect which cannot be solved by research.

Actors and stakeholders involving policy, enterprises, market institutions, and associations as representatives of enterprise or consumer groups need to come up with joint initiatives if transparency in the sector is to improve. The realization of transparency in the food sector cannot be solved by decisions of individual enterprises alone and depends on coordinated actions.

There are always closed groups, for example, individual retail groups with their suppliers, that could drive transparency forward. However, such initiatives are usually restricted to local or regional effects and have also met with resistance when suppliers refused to follow requests. Acknowledging the ownership of data will be an important element in reaching a sustainable information exchange. It is a prerequisite of cooperation.

The development of information markets is still at a low level. However, with the increasing interest in transparency, the information provided with products is an added value with market relevance between retail and consumers, but could also be between all members of the food value chain. If appropriately appreciated, cooperation will evolve.

To sum it up, moving forward in transparency needs research and action. It needs coordination in research to eliminate deficiencies in line with developments in transparency. It needs coordination in action to move forward and to assure that investments in individual enterprises match certain levels of transparency agreed upon and are initiated simultaneously to reap the benefits improvements in transparency could initiate.

It is common understanding throughout the industry that improvements in transparency are decisive elements in developments toward increased sustainability of the sector. This is demonstrated through industry initiatives like the Global Reporting Initiative (GRI) and others. However, all these initiatives are limited in scope and do not solve the inherent coordination problem.

It will remain to be seen if the sector can manage coordination itself or will depend on outside support including the involvement of policy through coordination initiatives (e.g., through the establishment of institutional support) or through regulatory interference that could drive developments forward.

Reference List (Reports to the European Commission)

The analysis in this book is based on the following list of reports which are comprised of more than 1500 pages of analysis. The reports involve the literature references relevant for the issues covered.

Public deliverables are open for distribution and available on the website of the project. Internal deliverables are not for distribution. Readers with interest in the themes should contact the authors directly. All reports are referenced as Project Report, EU Project Transparent_Food, European Commission, Brussels.

PUBLIC DELIVERABLES

Barling, D., Simpson, D., 2010. Report to the European Commission. Deliverable D5.1 Report drawn from data collection and review and stakeholder participant workshops on the breadth and range of certification systems and labelling schemes signalling information to consumers and the strengths and weaknesses of these systems and signals.

Barling, D., Simpson, D., 2010. Report to the European Commission. Deliverable D5.3 Report of analysis public policy statements and legal requirements and expectations of food transparency schemes and their role in achieving policy and regulatory goals.

Barling, D., Simpson, D., Jespersen, L., Halberg, N., Hermansen, J., 2010. Report to the European Commission. Deliverable D4.2 State of the art on information use in food chains with relevance for ethical and social concerns.

Barling, D., Chryssochoidis, G., Simpson, D., Kehagia, O., 2011. Report to the European Commission. Deliverable D5.4 Final summary report integrating report findings from Tasks 5.1–5.3.

Gellynck, X., Van Lembergen, K., Molnár, A., Sebök, A., Berczeli, A., 2011. Report to the European Commission. Deliverable D6.2 Analysis of selected experiences of "best practice transparency solutions" in enterprises and food chains.

Hermansen, J., Halberg, N., Jespersen, L., Östergren, K., Sonesson, U., Lorentzon, K., et al., 2011. Report to the European Commission. Deliverable D4.5 Specification of critical research needs and priorities with relevance for environmental, ethical and social concerns and for improvements in food chain transparency.

Martini, D., Mietzsch, E., Giannerini, G., Papaekonomou, V., Kunisch, M., 2010. Report to the European Commission. Deliverable D2.1 Specification of tracking/tracing requirements; analysis and evaluation of selected tracking/tracing solution alternatives.

Martini, D., Mietzsch, E., Kunisch, M., 2011. Report to the European Commission. Deliverable D2.2 Feasibility study taking into account different ways of serving tracking/tracing needs within the sector.

Östergren, K., Sonesson, U., Lorentzon, K., Hermansen, J., Jespersen, L., Halberg, N., 2010. Report to the European Commission. Deliverable D4.1 State of the art on information use in food chains with relevance for environmental concerns.

Schiefer, G., 2010. Report to the European Commission. Deliverable D7.2 Focus guide on transparency, tracking, tracing, sustainability and integrity.

Sebök, A., Berczeli, A., Homolka, F., 2011. Report to the European Commission. Deliverable D8.3 Report on the organization of stakeholder workshops/conferences and the development of transparency workshop material.

Sebök, A., Berczeli, A., Homolka, F., Molnar, A., Gellynck, X., Dégen, G., et al., 2011. Report to the European Commission. Deliverable D8.2 Report on the development of best practice guides and design of dissemination material for stakeholders.

INTERNAL DELIVERABLES

Chryssochoidis, G., Kehagia, O., Barling, D., Simpson, D., 2010. Report to the European Commission. Deliverable D5.2 A meta-analytic roadmap of consumer requests, contexts for such requests and priorities.

Gellynck, X., Van Lembergen, K., Molnár, A., Sebök, A., Berczeli, A., 2010. Report to the European Commission. Deliverable D6.1 Topics, performance indicators, and template for analysis of best practice experiences.

Gellynck, X., Van Lembergen, K., Molnár, A., Sebök, A., Berczeli, A., 2011. Report to the European Commission. Deliverable D6.3 Framework for the formulation of "best practice guides."

Hofstra, H., Hogg, T., Sebök, A., 2010. Report to the European Commission. Deliverable D3.1 Food safety assessment along the production chain. Impact of raw material status, transportation and storage as well as processing, packaging and distribution.

Knorr, D., Jäger, H., Surowsky, B., 2010. Report to the European Commission. Deliverable D3.2 Determination of food quality parameters of selected products according to sensorial and nutritional quality definitions.

Knorr, D., Hofstra, H., Hogg, T., Jäger, H., Surowsky, B., 2011. Report to the European Commission. Deliverable D3.5 Specification of critical research needs and priorities with relevance for food safety and quality concerns and for improvements in food chain transparency.

Knorr, D., Jäger, H., Surowsky, B., 2011. Report to the European Commission. Deliverable D3.4 Analysis of deficiencies (weaknesses) within traditional food processing, of improvement opportunities through process optimization or emerging technologies, and of feasibility regarding industrial implementation.

Knorr, D., Jäger, H., Surowsky, B., Hofstra, H., Hogg, T., 2011. Report to the European Commission. Deliverable D3.3 Analysis, evaluation, and documentation of selected 'best practice' monitoring and reporting schemes.

Martini, D., Mietzsch, E., Schmitz, M., Kunisch, M., 2011. Report to the European Commission. Deliverable D2.3 Formulation of a blueprint proposal for a European backbone solution.

Östergren, K., Sonesson, U., 2011. Report to the European Commission. Deliverable D4.4 Analysis of deficiencies (weaknesses) within traditional food processing, of improvement opportunities through process optimization or emerging technologies, and of feasibility regarding industrial implementation.

Östergren, K., Sonesson, U., Lorentzon, K., Hermansen, J., Halberg, N., 2011. Report to the European Commission. Deliverable D4.3 Analysis, evaluation, and documentation of selected "best practice" monitoring and reporting schemes.

Schiefer, G., 2010. Report to the European Commission. Deliverable D7.1 Framework for analysis, documentation and evaluation.

Sebök, A., Berczeli, A., 2010. Report to the European Commission. Deliverable D8.1 Report on dissemination plan and stakeholder groups for inclusion in project activities and dissemination of results.

Sebök, A., Berczeli, A., Homolka, F., 2011. Report to the European Commission. Deliverable D8.4 Report on dissemination activities.

Glossary of Selected Transparency Indicators

Carbon Footprint A measure of the total amount of *carbon dioxide* (CO_2) and *methane* (CH_4) emissions of a defined system or activity of interest. Calculated as carbon dioxide equivalent (CO_2e) using the relevant 100-year *global warming potential* (GWP100).

Certification Schemes Products produced within a certain certification scheme linked to products, production processes, or production sites provide a body of information represented in the certification rules and requirements laid down by the owners of the certification scheme. Examples involve the IFS (International Food Standard), BRC (British Retailer Consortium), or Global G.A.P standards.

Fair Trade The fair trade initiative is linked to an organization with a focus on better prices, decent working conditions, local sustainability, and fair terms of trade for farmers and workers in the developing world. It is complemented by a certification scheme.

Fair Miles The fair miles concept builds on the food miles concept complemented by social and economic development aspects.

Food Miles The concept of "food miles" presents an argument to buy goods which have traveled the shortest distance from farm to table, and to discriminate against long-haul transportation, especially air-freighted goods.

Global Warming Potential (GWP) GWP is a value used to compare the abilities of different greenhouse gases to trap heat in the atmosphere. GWPs are based on the heat-absorbing ability of each gas relative to that of carbon dioxide (CO_2), as well as the decay rate of each gas (the amount removed from the atmosphere over a given number of years).

Greenhouse Gases (GHGs) GHGs are gases that trap heat in the atmosphere. This process is the fundamental cause of the *greenhouse effect*. The primary GHGs in the *Earth's atmosphere* are *water vapor*, *carbon dioxide*, *methane*, *nitrous oxide*, and *ozone*.

Labels Labels represent a body of information that is either directly provided with the label or represented in the certification rules and requirements laid down by the organization that owns the label.

Water Footprint A water footprint consists of three components: the *blue*, *green*, and *gray water footprint*. The blue water footprint is the volume of freshwater that evaporated from the global blue water resources (surface water and ground water) to produce the goods and services consumed. The green water footprint is the volume of water evaporated from the global green water resources (rainwater stored in the soil as soil moisture). The gray water footprint is the volume of polluted water that associates with the production of goods and services.

Glossary of Selected Transparency Indicators

Carbon Footprint A measure of the total amount of *carbon dioxide* (CO_2) and *methane* (CH_4) emissions of a defined system or activity of interest. Calculated as carbon dioxide equivalent (CO_2e) using the relevant 100-year *global warming potential* (GWP100).

Certification Schemes Products produced within a certain certification scheme linked to products, production processes, or production sites provide a body of information represented in the certification rules and requirements laid down by the owners of the certification scheme. Examples involve the IFS (International Food Standard), BRC (British Retailer Consortium), or Global G.A.P standards.

Fair Trade The fair trade initiative is linked to an organization with a focus on better prices, decent working conditions, local sustainability, and fair terms of trade for farmers and workers in the developing world. It is complemented by a certification scheme.

Fair Miles The fair miles concept builds on the food miles concept complemented by social and economic development aspects.

Food Miles The concept of "food miles" presents an argument to buy goods which have traveled the shortest distance from farm to table, and to discriminate against long-haul transportation, especially air-freighted goods.

Global Warming Potential (GWP) GWP is a value used to compare the abilities of different greenhouse gases to trap heat in the atmosphere. GWPs are based on the heat-absorbing ability of each gas relative to that of carbon dioxide (CO_2), as well as the decay rate of each gas (the amount removed from the atmosphere over a given number of years).

Greenhouse Gases (GHGs) GHGs are gases that trap heat in the atmosphere. This process is the fundamental cause of the *greenhouse effect*. The primary GHGs in the *Earth's atmosphere* are *water vapor*, *carbon dioxide*, *methane*, *nitrous oxide*, and *ozone*.

Labels Labels represent a body of information that is either directly provided with the label or represented in the certification rules and requirements laid down by the organization that owns the label.

Water Footprint A water footprint consists of three components: the *blue*, *green*, and *gray water footprint*. The blue water footprint is the volume of freshwater that evaporated from the global blue water resources (surface water and ground water) to produce the goods and services consumed. The green water footprint is the volume of water evaporated from the global green water resources (rainwater stored in the soil as soil moisture). The gray water footprint is the volume of polluted water that associates with the production of goods and services.

www.ingramcontent.com/pod-product-compliance
Lightning Source LLC
Chambersburg PA
CBHW072151020426
42334CB00018B/1952
* 9 7 8 0 1 2 4 1 7 1 9 5 4 *